Exercises and Solutions in Lightweight Design

Andreas Öchsner

Exercises and Solutions in Lightweight Design

Springer

Andreas Öchsner
Faculty of Mechanical and Systems
Engineering
Esslingen University of Applied Sciences
Esslingen, Baden-Württemberg, Germany

ISBN 978-3-658-49332-5 ISBN 978-3-658-49333-2 (eBook)
https://doi.org/10.1007/978-3-658-49333-2

Translation from the German language edition: "Aufgaben und Lösungen zum Stoff- und Formleichtbau" by Andreas Öchsner, © Der/die Herausgeber bzw. der/die Autor(en), exklusiv lizenziert an Springer Fachmedien Wiesbaden GmbH, ein Teil von Springer Nature 2023. Published by Springer Fachmedien Wiesbaden. All Rights Reserved.

This Springer imprint is published by the registered company Springer Fachmedien Wiesbaden GmbH, part of Springer Nature.
The registered company address is: Abraham-Lincoln-Str. 46, 65189 Wiesbaden, Germany

Preface

Lightweight design concepts, such as material and shape-based lightweight design, can be understood as the application of classic engineering concepts and disciplines to reduce structural weight. In particular, basic knowledge of applied mechanics, materials science, manufacturing technology, and design theory is applied here. To introduce the various lightweight design concepts within the framework of a university engineering degree program, simple structural elements, which are presented in the context of applied mechanics, are recommended first. The simplest elements are bars and beams, which, along with springs, are classified as one-dimensional structural elements. Using these elements, questions regarding material selection and the geometric design and optimization of load-bearing structures can be discussed quite clearly. Based on these one-dimensional structural elements, a plenitude of hand calculation problems can be easily generated to effectively practice the various concepts.

This exercise book is intended as a supplement to my textbooks on lightweight design. However, the successful use of the textbooks in the Bachelor's degree course *Structural Optimization* and the English-language Master's lecture *Lightweight Design* has also shown that students have a very strong interest in exercises with detailed solutions. This exercise book is intended to meet this expectation.

Esslingen, Germany
July 2025

Andreas Öchsner

Competing Interests The author has no competing interests to declare that are relevant to the content of this manuscript.

Contents

Symbols and Abbreviations

Latin Symbols (Capital Letters)

A	Area, cross-sectional area
A_s	Shear area
AG	Shear stiffness
B_1	Factor
B	Factor
C	Constant
E	Young's modulus
$\overline{EA}$	Average tensile stiffness
$\overline{EI}_y$	Average bending stiffness
F	Force, Yield condition
F_G	Dead weight
F_{cr}	Critical force (buckling force)
G	Shear modulus
$\mathcal{H}$	Axial first moment of area
I	Axial second moment of area
I_p	Polar second moment of area
K	Buckling coefficient
L	Length
M	Lightweight index, Moment
N	Normal force
Q	Shear force
$R_{p0.2}$	Initial yield stress
SEA	Specific energy absorption
V	Volume

Latin Symbols (Small Letters)

a	Geometric dimension
b	Geometric dimension

c	Damper constant, Constant of integration
d	Diameter
f	Function
g	Gravitational acceleration, Function
h	Geometric dimension
r_c	Average sandwich thickness
Δh	Layer thickness (sandwich)
Δh^F	Face sheet thickness (sandwich)
$\Delta h^{F,n}$	Length specific face sheet thickness (sandwich)
Δh^C	Core thickness (Sandwich)
$\Delta h^{C,n}$	Length specific core thickness (sandwich)
k_s	Spring constant, Geometric and material parameter, Shear yield stress, Shear correction factor
k_t	Tensile yield stress
k_t^{init}	Initial tensile yield stress
m	Mass, Length specific moment
m^n	Length specific mass
p	Distributed load in x-direction
q	Distributed load in z-direction
r	Radius
s	Geometric dimension
t	Geometric dimension
u	Displacement
w	Geometric dimension
x	Cartesian coordinate
y	Cartesian coordinate
z	Cartesian coordinate

Greek Symbols (Capital Letters)

Θ	Argument ($\Theta = \frac{2\pi \Delta h^C}{\lambda}$)
Π	Strain energy

Greek Symbols (Small Letters)

α	Parameter, Angle
γ	Shear strain
γ_{aB}	Shear strain at failure
ε	Normal strain
ε_A	Strain at failure
ε_{A_t}	Total strain at failure (including elastic part)
$\varepsilon_{p0.2_t}$	0.2-% strain limit (corresponding to $R_{p0.2}$)

κ	Curvature
λ	Parameter, Wave length
ν	Poisson's ratio
ϱ	Density
π	Volume specific strain energy
π^{el}	Elastic part of volume specific strain energy
π^{pl}	Plastic part of volume specific strain energy
π^{s}	Deviatoric part of the volume specific strain energy
π°	Hydrostatic part of the volume specific strain energy
σ	Stress, normal stress
σ_{cr}	Critical stress
σ_{eff}	Equivalent stress (effective stress)
σ_i	Principal stress ($i = 1, 2, 3$)
τ	Shear stress
τ_{aB}	Shear strength
τ_p	Shear yield stress
ϕ	Rotation angle, rotation
ϕ_F	Fiber volume fraction
φ	Rotation angle, Rotation, Angle of torsion

Mathematical Symbols

$\times$	Multiplication sign

Indices, Superscripted

$\ldots^{C}$	Core
$\ldots^{c}$	Compression
$\ldots^{E}$	Euler
$\ldots^{el}$	Elastic
$\ldots^{F}$	Face sheet
$\ldots^{k}$	Layer index
$\ldots^{pl}$	Plastic
$\ldots^{t}$	Tension

Indices, Subscripted

$\ldots_{o}$	Outer
$\ldots_{b}$	Bending
$\ldots_{F}$	Face sheet

$\ldots_{i}$	Internal
$\ldots_{C}$	Core
$\ldots_{m}$	Average value
$\ldots_{max}$	Maximum value
$\ldots_{ref}$	Reference
$\ldots_{s}$	Shear
$\ldots_{t}$	Torsion, Tension
$\ldots_{c}$	Center

Abbreviations

1D	One-dimensional
CFRP	Carbon fiber-reinforced plastic
UD	Unidirectional

1 Introduction and Motivation

Abstract

This chapter briefly explains the motivation for this workbook and provides some technical information.

This book is designed as a supplement to the series of textbooks on lightweight design concepts (see Öchsner (2018, 2019, 2020, 2022, 2025)). The concept of these textbooks,i.e., introducing the most important concepts of lightweight design starting with one-dimensional structural elements, has proven successful in two lectures over several years. The use of one-dimensional structural elements, i.e., bars and beams, allows students to practice the theory using classic hand calculation problems. Although the textbooks listed above contain a plenitude of exercises, the solutions are presented only in shortened form, i.e., with only essential results. This additional book is in line with students' demands for the most detailed solutions to the exercises possible. This exercise book is intended to provide the classic hand calculation problems and detailed solutions.

In the exercise book, references to the theory are made in some places, for example, in the form of equation numbers, in the textbook version Öchsner (2025). The corresponding reference numbers are then given the index „TB" to indicate that the corresponding reference refers to the textbook:

$$(\ldots)_{\mathrm{TB}}$$

Many exercises involve numerical calculations for steel, aluminum, and titanium (or corresponding alloys). The necessary material parameters are summarized in Table 1.1. If intervals were given in the corresponding reference, a mean value is also given in the table to simplify the calculations. Therefore, with the given solutions, which always refer to the mean value, it should be noted that trends can also change when moving to the upper or lower range of the interval.

A. Öchsner, *Exercises and Solutions in Lightweight Design*,
https://doi.org/10.1007/978-3-658-49333-2_1

Table 1.1 Properties of different materials: E: Young's modulus; ν: Poisson's ratio; ϱ: mass density; $R_{p0.2}$: initial yield stress; ε_A: strain at failure. The intervals are taken from Ashby und Jones (2005)

	E, $\frac{\text{N}}{\text{mm}^2}$	ν, –	ϱ, 10^{-6} $\frac{\text{kg}}{\text{mm}^3}$	$R_{p0.2}$, $\frac{\text{N}}{\text{mm}^2}$	ε_A, –
Stainless steel (austenitic)	195000	0.30	7.8	393	0.55
	190000 ... 200000		7.5 ... 8.1	286 ... 500	0.45 ... 0.65
Al alloys	74000	0.33	2.75	364	0.175
	69000 ... 79000		2.6 ... 2.9	100 ... 627	0.05 ... 0.3
Ti alloys	105000	0.35	4.7	750	0.18
	80000 ... 130000		4.3 ... 5.1	180 ... 1320	0.06 ... 0.3

References

Ashby, M.F., Jones, D.R.H.: Engineering Materials 1: An Introduction to Properties, Applications and Design. Elsevier, Amsterdam (2005)

Öchsner, A.: Leichtbaukonzepte: Eine Einführung anhand einfacher Strukturelemente für Studierende. Springer Vieweg, Wiesbaden (2018)

Öchsner, A.: Leichtbaukonzepte anhand einfacher Strukturelemente: Neuer didaktischer Ansatz mit zahlreichen Übungsaufgaben. Springer Vieweg, Wiesbaden (2019)

Öchsner, A.: Stoff- und Formleichtbau: Leichter Einstieg mit eindimensionalenStrukturen. Springer Vieweg, Wiesbaden (2020)

Öchsner, A.: Stoff- und Formleichtbau: Leichter Einstieg mit eindimensionalenStrukturen. Springer Vieweg, Wiesbaden (2022)

Öchsner, A.: Lightweight Design: An Introduction Based on One-Dimensional Structures. Springer Vieweg, Wiesbaden (2025)

Exercises 2

Abstract

This chapter contains exercises on the topics of mechanics of materials, lightweight design in regard to the material selection, shape optimization and sandwich elements, with the subchapters strength analysis, limit loading and optimization.

2.1 Basics of Mechanics of Materials

2.1.1 Simplified model of a tower under dead weight
Given is a simplified model of a tower which is deforming under the influence of its dead weight, cf. Fig. 2.1. The tower is of original length L, cross-sectional area A, Young's modulus E, and mass density ϱ. The standard gravity is given by g.
Calculate:

(a) The stress distribution $\sigma_x(x)$ in the member.
(b) The reduced length $L' = L - u_x(L)$ due to the acting dead weight.
(c) The maximum length L_{max} if a given stress limit σ_{max} at the foundation ($x = 0$) cannot be exceeded.

2.1.2 Course of the bending line and maximum stress for a cantilever beam with different loads
For the cantilever beam in Fig. 2.2, calculate and sketch the course of the bending line and the maximum stress (σ_x) along the longitudinal axis for both load cases according to the Euler-Bernoulli theory. The geometric dimensions (I, L) and the material parameter (E) are assumed to be given and constant. For simplification, a square cross-section (side length: a) can also be assumed.

2.1.3 Course of bending line, stress and strain for a bi-material cantilever beam
For the bi-material cantilever beam in Fig. 2.3, calculate the course of the bending

A. Öchsner, *Exercises and Solutions in Lightweight Design*,
https://doi.org/10.1007/978-3-658-49333-2_2

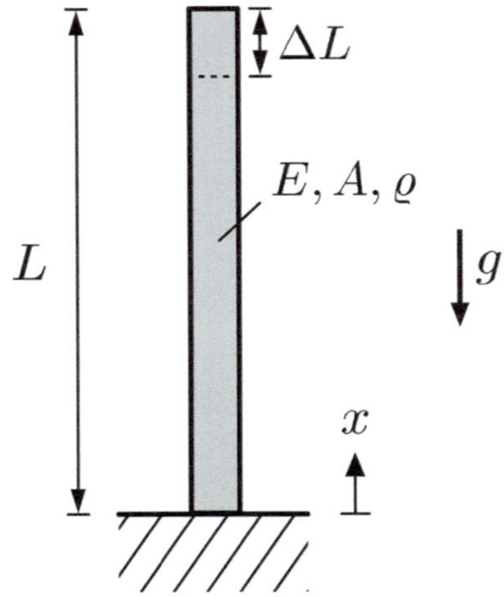

Fig. 2.1 Simplified model of a tower under its dead weight

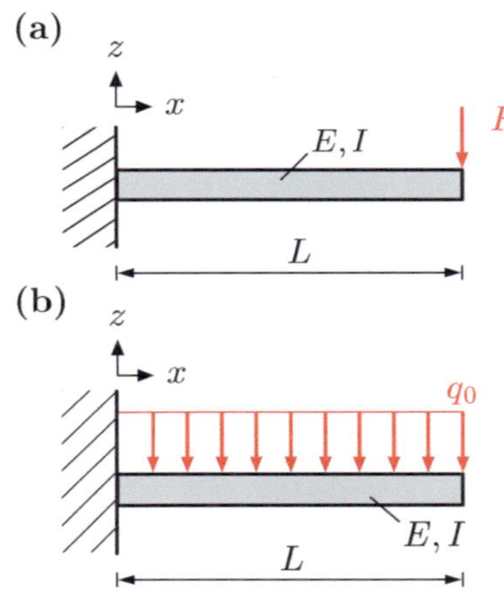

Fig. 2.2 Cantilever beam loaded by **a** a single force F_0 and **b** a constant distributed load q_0

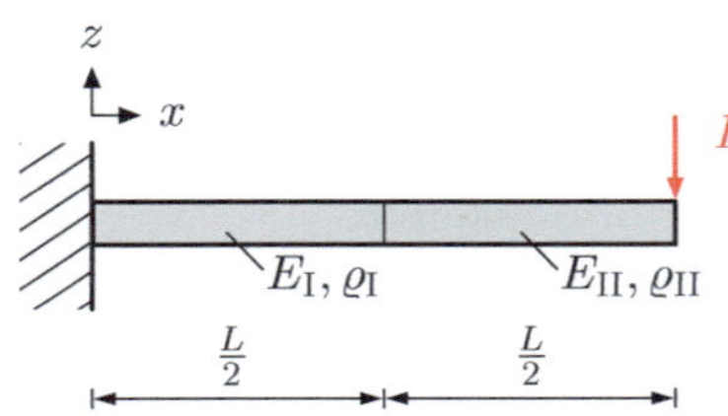

Fig. 2.3 Bi-material cantilever beam loaded by a single force F_0

line, the stress and strain on the top surface along the longitudinal axis according to the Euler-Bernoulli theory. The geometric dimensions (I, L) and the two material parameters (E_{I} and E_{II}) are assumed to be given and constant. For simplification, a square cross-section (side length: a) can also be assumed. To graphically represent the three courses, $E_{\text{I}} = 2E_{\text{II}}$ can be set.

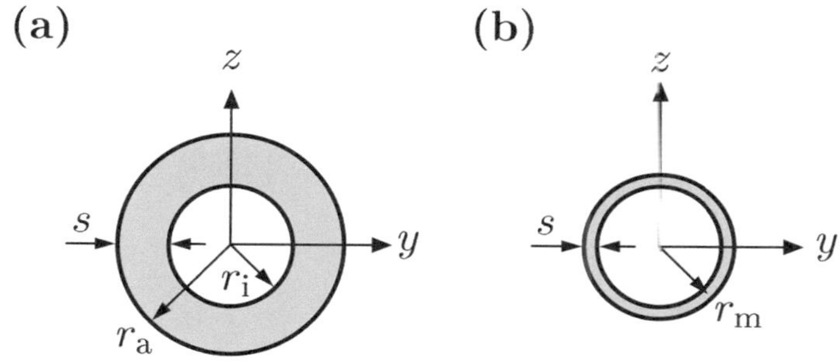

Fig. 2.4 Second moment of area for a circular ring: **a** thick and **b** thin wall thickness

2.1.4 Comparison of the maximum deflections of a cantilever beam with constant line load according to different beam theories

For a cantilever beam with constant line load q_0 (rectangular cross-section with width b and height h, beam length L, q_0 in negative z-direction), calculate the normalized maximum deflection according to the theories of (a) Euler-Bernoulli, (b) Timoshenko ($k_s = 5/6$) and (c) Levinson as a function of the slenderness ratio h/L. The maximum deflection according to Euler-Bernoulli should be used as the normalization factor.

2.1.5 Axial second moment of area for a circular ring

Calculate the axial second moment of area I for a circular ring, first for a thick wall thickness s (see Fig. 2.4a). Then simplify the result for a thin circular ring, i.e. $s \ll r_m$ (see Fig. 2.4b). For which ratio of $\frac{r_m}{s}$ is the relative error of the simplification smaller than 1%?

Finally, corresponding equations for the cross-sectional area A should be given.

2.1.6 Axial second moment of area for a square box profile

Calculate the axial second moment of area I for a square box profile, first for a thick wall thickness s (see Fig. 2.5a). Then simplify the result for a thin box profile, i.e. $s \ll a_m$ (see Fig. 2.5b). For which ratio of $\frac{a_m}{s}$ is the relative error of the simplification smaller than 1%?

Finally, corresponding equations for the cross-sectional area A should be given.

2.1.7 General solution of the differential equations of the Timoshenko beam using a computer algebra systems

The coupled differential equations of the Timoshenko beam according to Eq. (2.19)$_{TB}$–(2.20)$_{TB}$ can be solved quite easily using computer algebra systems. By using Maple® one obtains the following representation:

$$\mathrm{u}(x) = \frac{qx^4}{24EI} + \frac{C1x^3}{6} + \frac{C2x^2}{2} + C3x + C4\,, \tag{2.1}$$

$$\phi(x) = -\frac{qx^3}{6EI} - \frac{C1x^2}{2} - C2x - C3 - \frac{qx}{GAK} - \frac{EIC1}{GAK}\,. \tag{2.2}$$

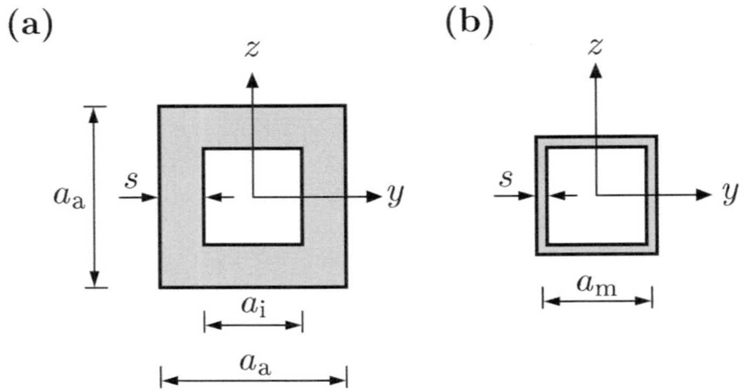

Fig. 2.5 Second moment of area for a squared box section: **a** thick and **b** thin wall thickness

However, a calculation using Maxima yields the following representation ($m = 0$):

$$\begin{aligned}
\mathrm{u}(x) = & -\frac{x^3\left(kAG\left(\frac{\mathrm{d}}{\mathrm{d}x}\mathrm{u}(x)\big|_{x=0}\right)+\phi(0)kAG\right)}{24EI}-\frac{kAGx^3\left(\frac{\mathrm{d}}{\mathrm{d}x}\mathrm{u}(x)\big|_{x=0}\right)}{8EI}\\
& + x\left(\frac{\mathrm{d}}{\mathrm{d}x}\mathrm{u}(x)\bigg|_{x=0}\right)-\frac{x^2\left(2kAG\left(\frac{\mathrm{d}}{\mathrm{d}x}\phi(x)\big|_{x=0}\right)+2q\right)}{24kAG}\\
& -\frac{x^2\left(kAG\left(\frac{\mathrm{d}}{\mathrm{d}x}\phi(x)\big|_{x=0}\right)+q\right)}{6kAG}-\frac{x^2\left(\frac{\mathrm{d}}{\mathrm{d}x}\phi(x)\big|_{x=0}\right)}{4}\\
& +\frac{q\,x^4}{24EI}-\frac{\phi(0)kAGx^3}{8EI}-\frac{qx^2}{4kAG}+\mathrm{u}(0),
\end{aligned} \tag{2.3}$$

$$\begin{aligned}
\phi(x) = & \frac{x^2\left(kAG\left(\frac{\mathrm{d}}{\mathrm{d}x}\mathrm{u}(x)\big|_{x=0}\right)+\phi(0)kAG\right)}{6EI}+\frac{kAGx^2\left(\frac{\mathrm{d}}{\mathrm{d}x}\mathrm{u}(x)\big|_{x=0}\right)}{3EI}\\
& + x\left(\frac{\mathrm{d}}{\mathrm{d}x}\phi(x)\bigg|_{x=0}\right)-\frac{qx^3}{6EI}+\frac{\phi(0)kAGx^2}{3EI}+\phi(0)\,.
\end{aligned} \tag{2.4}$$

Show that both representations can be transformed into each other.

2.1.8 Maximum difference between von Mises and Tresca

Calculate the maximum absolute error between the equivalent stress hypotheses according to von Mises and Tresca in the σ-τ stress plane as a percentage of k_t. Furthermore, state the maximum relative error.

2.1.9 Bar under tensile and torsional loading

For the bar shown in Fig. 2.6 (length L, cross-sectional area A, elastic modulus E, and shear modulus G), calculate the maximum equivalent stress according to von Mises. The bar is loaded by a single force F_0 and by a torsional moment M_t.

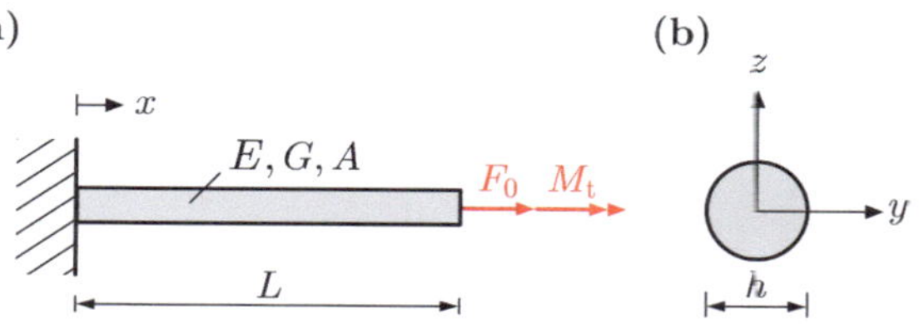

Fig. 2.6 Bar under tensile and torsional loading: **a** general configuration and **b** cross-sectional profile

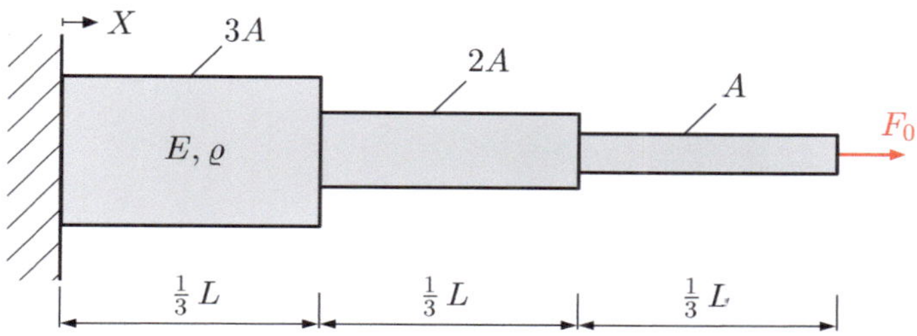

Fig. 2.7 Stepped tensile bar

2.2 Lightweight Design Concepts: Material Selection

2.2.10 Finite element calculation of a stepped tensile bar

Calculate the lightweight index M for the tensile bar shown in Fig. 2.7 using a finite element approach. For this purpose, three linear bar elements of length $\frac{L}{3}$ are to be used. The entire bar is made of a homogeneous material with density ϱ and elastic modulus E.

2.2.11 Calculation of the lightweight index for a cantilever beam with constant distributed load

Calculate for the beam shown in Fig. 2.8 the general lightweight index for stainless steel (St), Al and Ti alloys under consideration of a stress criterion. The material properties can be taken from Table 1.1 as average values. Furthermore, assume for the calculation of the numerical values a constant geometry ($L = 100$ mm, $b = h = 10$ mm).

2.2.12 Calculation of the lightweight index for a simply supported beam with constant distributed load

Calculate for the beam shown in Fig. 2.9 the general lightweight index for stainless steel (St), Al and Ti alloys under consideration of a stress criterion. The material properties can be taken from Table 1.1 as average values. Furthermore, assume for the calculation of the numerical values a constant geometry ($L = 100$ mm, $b = h = 10$ mm).

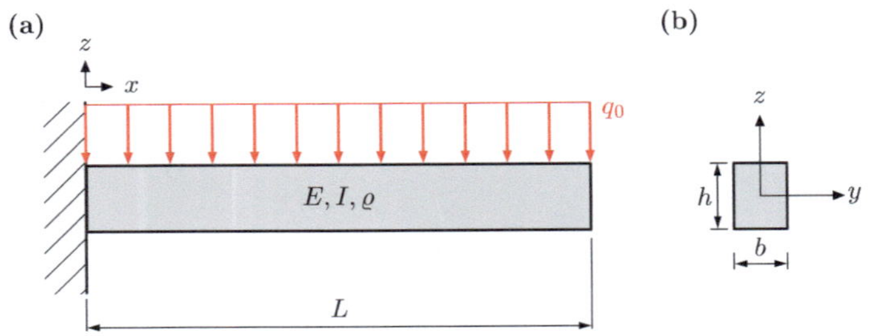

Fig. 2.8 Cantilever beam with constant distributed load: **a** general configuration and **b** cross-sectional profile

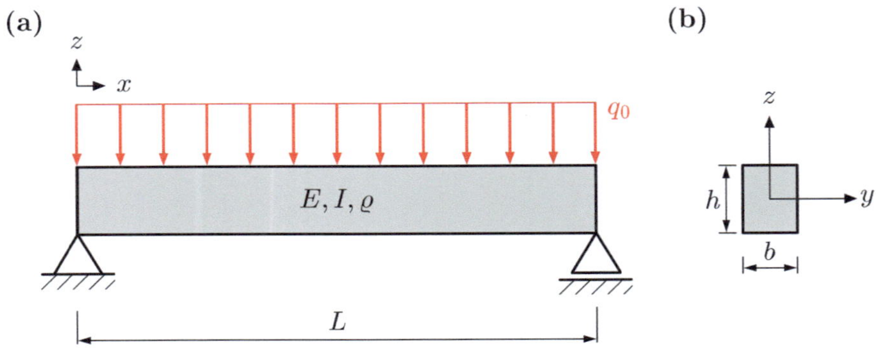

Fig. 2.9 Simply supported beam with constant distributed load: **a** general configuration; **b** cross section

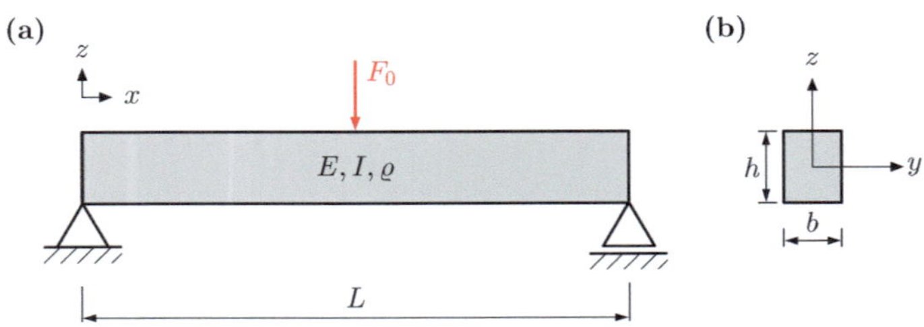

Fig. 2.10 Simply supported beam with point load: **a** general configuration; **b** cross section

2.2.13 Calculation of the lightweight index for a simply supported beam with point load

Calculate for the beam shown in Fig. 2.10 the general lightweight index for stainless steel (St), Al and Ti alloys under consideration of a stress criterion. The material properties can be taken from Table 1.1 as average values. Furthermore, assume for the calculation of the numerical values a constant geometry ($L = 100$ mm, $b = h = 10$ mm).

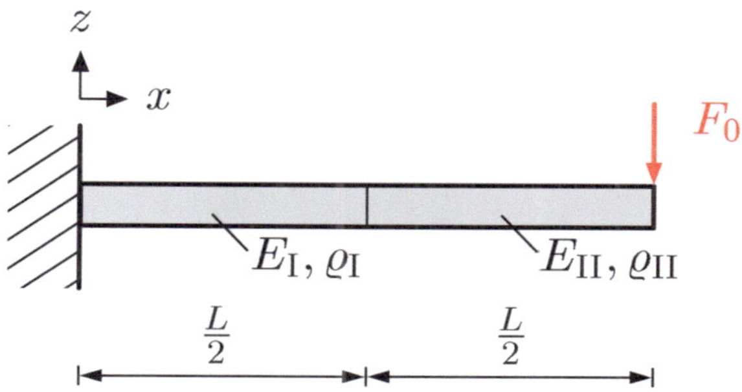

Fig. 2.11 Bi-material cantilever beam loaded by a single force F_0

2.2.14 Calculation of the specific energy absorption for a cantilever beam with constant distributed load

Calculate for Problem 2.2.11 the specific energy absorption.

2.2.15 Calculation of the specific energy absorption for a simply supported beam with constant distributed load

Calculate for Problem 2.2.12 the specific energy absorption.

2.2.16 Calculation of the specific energy absorption for a simply supported beam with point load

Calculate for Problem 2.2.13 the specific energy absorption.

2.2.17 Calculation of specific energy absorption for a bi-material cantilever beam

Calculate for the bi-material cantilever in Fig. 2.11 the specific energy absorption for a given force F_0. The geometric dimensions (I, L) and the material parameters (E_{I}, ϱ_{I} and E_{II}, ϱ_{II}) are assumed to be given. For simplification, a square cross-section (side length: a) can also be assumed.

2.3 Lightweight Design Concepts: Shape Optimization

2.3.18 Analysis of the lightweight index for a cantilever beam with end force and tubular cross section

For the cantilever beam with end force shown in Fig. 2.12, the lightweight index M is to be analyzed based on a stress criterion. To do this, first derive the general equation of the lightweight index as a function of d and s, i.e. $M = M(d, s)$. Then, the functions $F_0(d, s)$, $F_{\mathrm{G}}(d, s)$ and $M(d, s) = F_0(d, s)/F_{\mathrm{G}}(d, s)$ are to be sketched as a function of (a) s at $d = 10\,\text{mm} = \text{const.}$ and of (b) d at $s = 2\,\text{mm} = \text{const.}$

2.3.19 Implementation of the concept of lightweight design shapes: Optimization of a box profile

For a cantilever beam ($L = 100\,\text{mm}$) with end force F_0, the cross section is to be optimized based on a stress criterion using the lightweight index. The original square

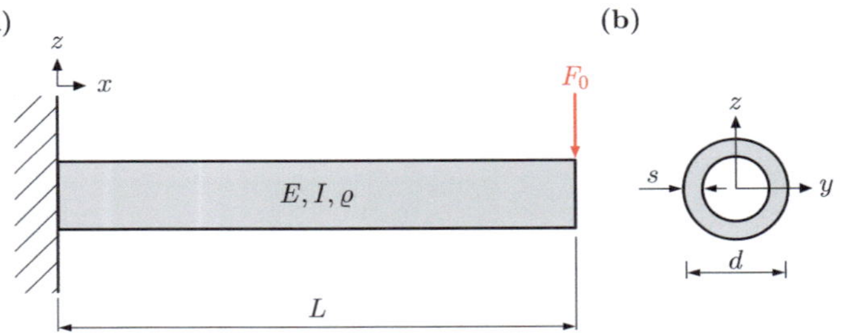

Fig. 2.12 Cantilever beam with point load: **a** General configuration; **b** beam cross section

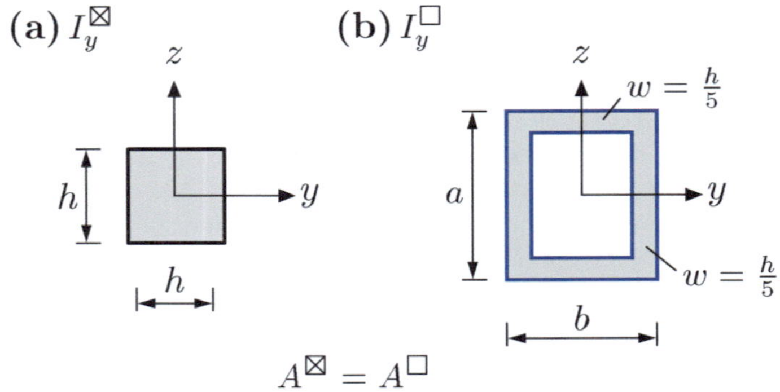

Fig. 2.13 Realization of the concept of lightweight design by shape optimization based on $I_y^{\square} > I_y^{\boxtimes}$ and $A^{\boxtimes} = A^{\square}$: **a** Original cross section with $b = h$; **b** modified cross section

cross-section ($h = 10\,\text{mm}$) is to be modified and optimized by a box profile while maintaining the beam weight (i.e. with $A^{\boxtimes} = A^{\square}$), see Fig. 2.13.

The optimization is to be carried out for stainless steel with the material properties according to Table 1.1.

2.3.20 Optimization of a beam in bending along its axis

For a cantilevered beam in bending (bending stiffness $E \times I(x)$, length L) with a single force F_0 at the free end, optimize the tubular cross section along the beam axis x taking into account a stress criterion (0.2-% strain limit $R_{p0.2}$). The cross-sectional dimensions at the clamping point $x = 0$ (see Fig. 2.14a) are given by the outside diameter d_0 and the wall thickness $s_0 = \frac{d_0}{5}$. Determine:

(a) the beam contour $d = d(x)$ along the beam axis (including sketch),
(b) the ratio of the masses of the optimized cross section with a beam of constant cross section according to Fig. 2.14a,
(c) the ratio of the lightweight index of the optimized cross section with a beam of constant cross section according to Fig. 2.14a,
(d) the ratio of the specific energy absorption of the optimized cross section with a beam of constant cross section according to Fig. 2.14a.

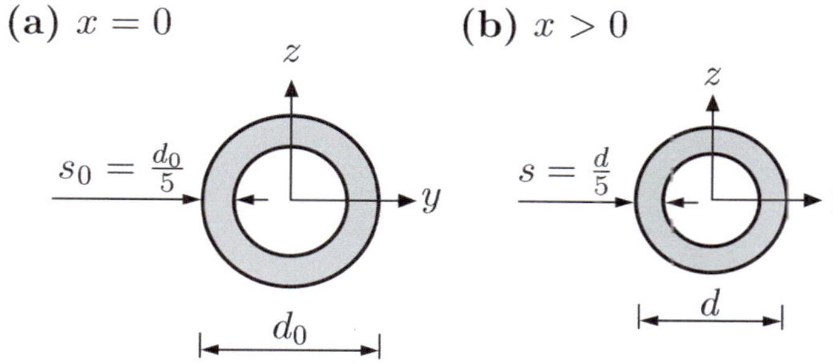

Fig. 2.14 Beam with changing cross section along the axis: **a** location of support; **b** arbitrary position

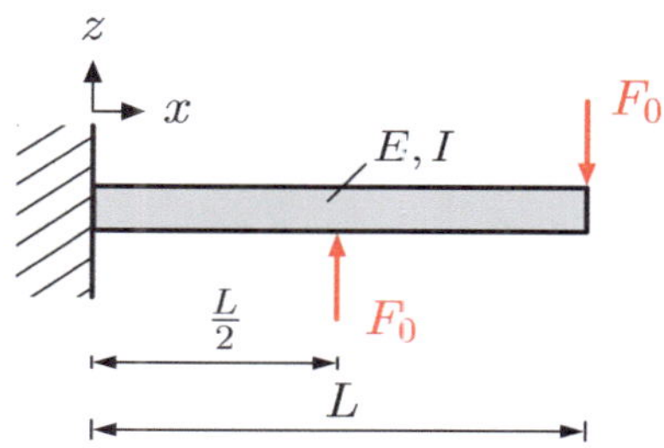

Fig. 2.15 Beam loaded with two single forces (same magnitude)

2.3.21 Optimization of a cantilevered beam loaded with two single forces of the same magnitude

For the cantilevered bending beam shown in Fig. 2.15 (bending stiffness $E \times I(x)$, length L) with two single forces F_0, optimize the tubular cross section along the beam axis x taking into account a stress criterion (0.2-% strain limit $R_{p0.2}$). The cross-sectional dimensions at the clamping point $x = 0$ (see Fig. 2.14a) are given by the outside diameter d_0 and the wall thickness $s_0 = \frac{d_0}{5}$. The cross-sectional ratios at any location x are shown in Fig. 2.14b. Determine:

(a) the beam contour $d = d(x)$ along the beam axis (including sketch),
(b) the specific energy absorption SEA of the optimized cross section,
(c) the problem in considering the lightweight index M in this case.

2.3.22 Optimization of a beam under 3-point bending: tubular cross section

For a beam under 3-point bending (bending stiffness $E \times I(x)$, length L, see Fig. 2.16a) which is loaded by a central single force F_0, optimize the tubular cross section along the beam axis x taking into account a stress criterion (0.2-% strain limit $R_{p0.2}$), see Fig. 2.16b.

The cross-sectional dimensions in the middle at $x = L/2$ are given by the outer diameter d_0 and the wall thickness $s_0 = d_0/5$, and the cross-sectional ratios at any point $x \neq L/2$ are modified in the same ratio, see Fig. 2.16b. Determine:

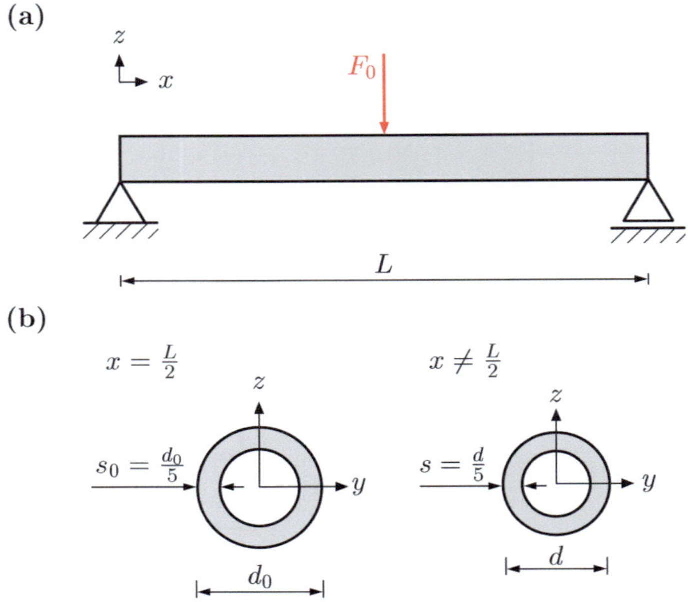

Fig. 2.16 Tubular beam under 3-point bending: **a** General configuration and **b** cross sections

(a) the beam contour $d = d(x)$ along the beam axis (including sketch),
(b) the specific energy absorption SEA of the optimized cross section. Compare this value with the specific energy absorption for a beam with a constant diameter d_0,
(c) the lightweight index M of the optimized cross section. Compare this value with the lightweight index for a beam with a constant diameter d_0,
(d) the problem of the technical realization of the contour $d = d(x)$.

2.3.23 Optimization of a beam under asymmetric 3-point bending

For a beam subjected to asymmetric 3-point bending (bending stiffness $E \times I(x)$, length L, see Fig. 2.17a) which is loaded by a single force F_0 at $x = b$, optimize the rectangular cross section along the beam axis x taking into account a stress criterion (0.2-% strain limit $R_{p0.2}$), see Fig. 2.17b.

The cross-sectional dimensions at $x = b$ are given by the outer diameter d_0 and the wall thickness $s_0 = d_0/6$, and the cross-sectional ratios at any point $x \neq b$ are modified in the same ratio, see Fig. 2.17b. Determine:

(a) the beam contour $d = d(x)$ along the beam axis (including sketch),
(b) the lightweight index M of the optimized cross section. Compare this value with the lightweight index for a beam with a constant diameter d_0,
(c) the specific energy absorption SEA of the optimized cross section. Compare this value with the specific energy absorption for a beam with a constant diameter d_0,
(d) the problem of the technical realization of the contour $d = d(x)$.

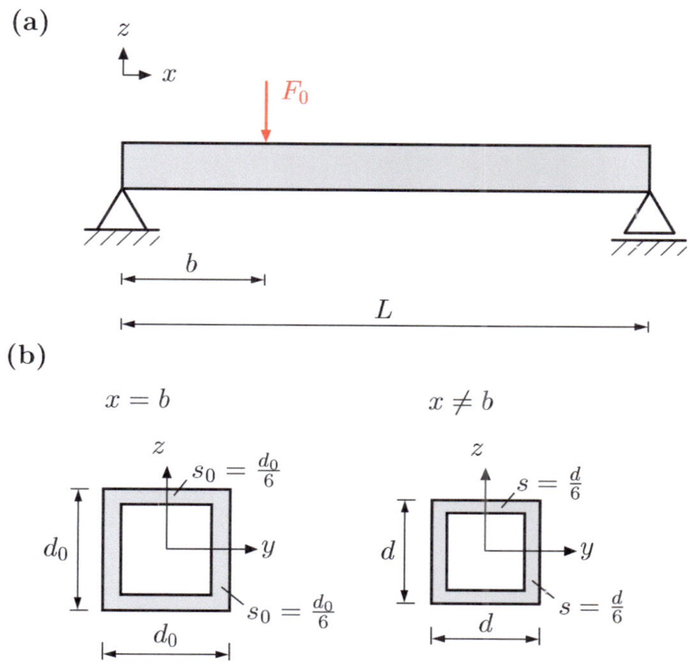

Fig. 2.17 Beam under asymmetrical 3-point bending: **a** General configuration and **b** cross sections

Special values must be specified for $b = \frac{1}{4}L$, $\frac{1}{3}L$ and $\frac{1}{2}L$.

2.3.24 Optimization of a simply supported beam under constant distributed load: rectangular cross section

For a simply supported beam (bending stiffness $E \times I(x)$, length L, see Fig. 2.18a) which is loaded by a constant distributed load q_0, optimize the rectangular cross section along the beam axis x taking into account a stress criterion (0.2-% strain limit $R_{p0.2}$), see Fig. 2.18b.

The cross-sectional dimensions in the middle at $x = L/2$ are given by the height h_0 and the constant width b, and the cross-sectional ratios at any point $x \neq L/2$ are given by the height $h(x)$ and the constant width b, see Fig. 2.18b. Determine:

(a) the optimized beam contour $h = h(x)$ along the beam axis x (including sketch),
(b) the specific energy absorption SEA of the optimized cross section ($h(x) \times b$). Compare this value with the specific energy absorption for a beam with a constant cross section ($h_0 \times b$),
(c) the problem of the technical realization of the contour $h = h(x)$.

Consider:

$\int_0^L \left(-\left[\frac{x}{L}\right]^2 + \left[\frac{x}{L}\right]^1\right)^{\frac{1}{2}} \mathrm{d}x = \frac{\pi L}{8}$ or $\int_0^L \left(-x^2 + Lx^1\right)^{\frac{1}{2}} \mathrm{d}x = \frac{\pi L^2}{8}$.

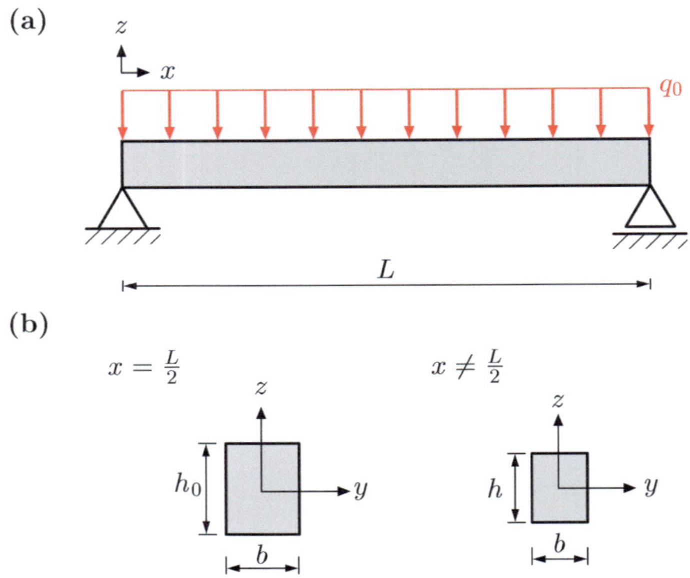

Fig. 2.18 Optimization of a simply supported beam under constant distributed load: **a** General configuration and **b** rectangular cross sections

2.3.25 Optimization of a simply supported beam under constant distributed load: tubular cross section (numerical integration)

For a simply supported beam (bending stiffness $E \times I(x)$, length L, see Fig. 2.19a) which is loaded by a constant distributed load q_0, optimize the tubular cross section along the beam axis x taking into account a stress criterion (0.2-% strain limit $R_{p0.2}$), see Fig. 2.19b.

The cross-sectional dimensions in the middle at $x = L/2$ are given by the outer diameter d_0 and the wall thickness $s_0 = d_0/6$, and the cross-sectional ratios at any point $x \neq L/2$ are modified in the same ratio, see Fig. 2.19b. Given are:

- Geometric dimension: $L = 2540\,\text{mm}$.
- Material properties of the beam: $E = 68948\,\text{MPa}$, $\varrho = 2691\,\text{kg/m}^3$, $R_{p0.2} = 247\,\text{MPa}$.
- Load: $q_0 = 1.05\,\text{N/mm}$.

Determine:

(a) the beam contour $d = d(x)$ along the beam axis (including sketch),
(b) the specific energy absorption SEA of the optimized cross section. Compare this value with the specific energy absorption for a beam with a constant diameter d_0,
(c) the problem of the technical realization of the contour $d = d(x)$.

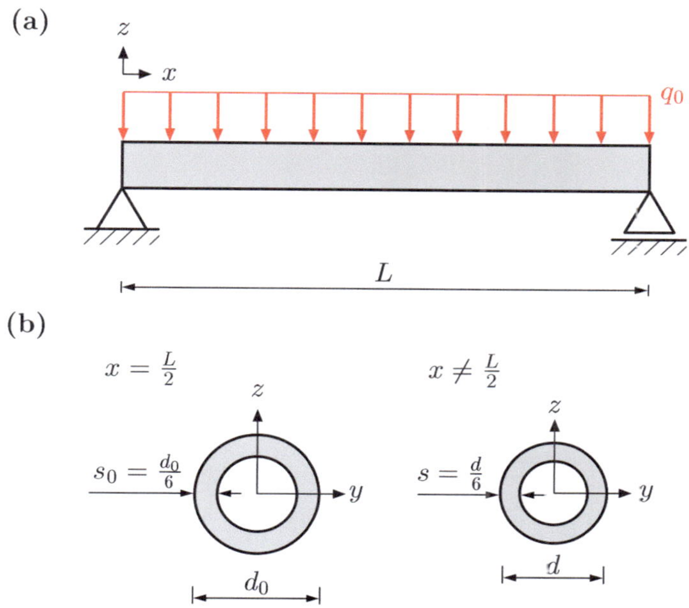

Fig. 2.19 Optimization of a simply supported beam under constant distributed load: **a** General configuration and **b** tubular cross sections

2.3.26 Optimization of a beam with a square box profile along the longitudinal axis under the influence of a constant distributed load

For a simply supported beam (bending stiffness $E \times I(x)$, length L, see Fig. 2.20a) which is loaded by a constant distributed load q_0, optimize the rectangular cross section along the beam axis x taking into account a stress criterion (0.2-% strain limit $R_{p0.2}$), see Fig. 2.20b. The cross-sectional dimensions in the middle at $x = L/2$ are given by the height d_0, the width d_0 and the constant thickness $s_0 = d_0/6$, and the cross-sectional ratios at any point $x \neq L/2$ are given by the height $d(x)$, the width $d(x)$ and the thickness $s(x) = d(x)/6$, see Fig. 2.20b. *Consider the comments on integration at the end of the problem!*
Determine:

(a) The optimized beam contour $d = d(x)$ along the beam axis x (including sketch).
(b) An approximate value for the lightweight index M_0 of the cross section with constant values (d_0, d_0, s_0).
(c) An approximate value for the lightweight index M of the optimized cross section $(d(x), d(x), s(x))$.
(d) The specific energy absorption SEA_0 of the cross section with constant values (d_0, d_0, s_0).
(e) The specific energy absorption SEA of the optimized cross section $(d(x), d(x), s(x))$.
(f) What can be said about the technical realization of the contour $d = d(x)$?

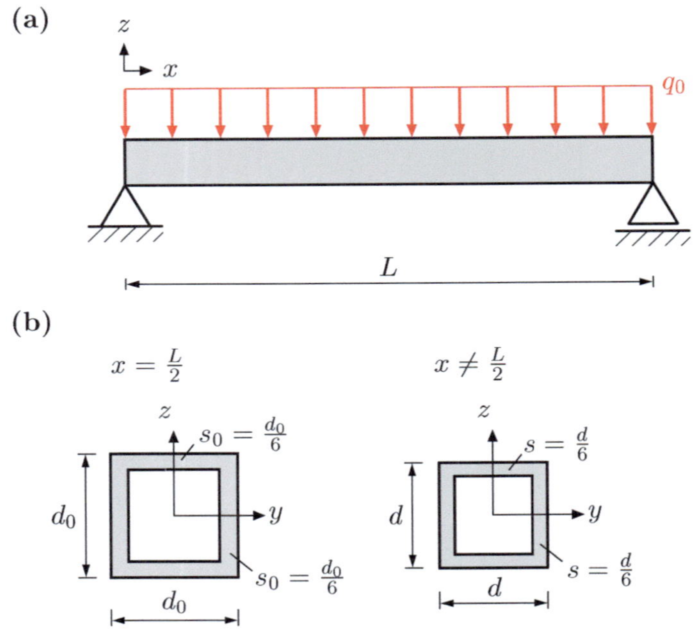

Fig. 2.20 Optimization of a simply supported beam under constant distributed load: **a** General configuration and **b** box-like cross sections

Remarks:
$\int_0^L (-(x/L)^2 + x/L)^2 \mathrm{d}x \approx L/30$
$\int_0^L (-(x/L)^2 + x/L)^{2/3} \mathrm{d}x \approx 0.293342L.$

2.3.27 Optimization of a beam with a rectangular box profile along the longitudinal axis under the influence of a constant distributed load
For a simply supported beam (bending stiffness $E \times I(x)$, length L, see Fig. 2.21a) which is loaded by a constant distributed load q_0, optimize the rectangular cross section along the beam axis x taking into account a stress criterion (0.2-% strain limit $R_{p0.2}$), see Fig. 2.21b. The cross-sectional dimensions in the middle at $x = L/2$ are given by the outer dimensions $2a_0 \times a_0$ and the constant thickness $d_0 = a_0/6$, and the cross-sectional ratios at any point $x \neq L/2$ are given by the outer dimensions $2a(x) \times a(x)$ and the thickness $d(x) = a(x)/6$, see Fig. 2.21b. *Consider the comments on integration at the end of the problem!*
Determine:

(a) The optimized beam contour $a = a(x)$ along the beam axis x (including sketch).
(b) An approximate value for the lightweight index M_0 of the cross section with constant value $(2a_0, a_0, d_0)$.
(c) An approximate value for the lightweight index M of the optimized cross section $(2a(x), a(x), d(x))$.

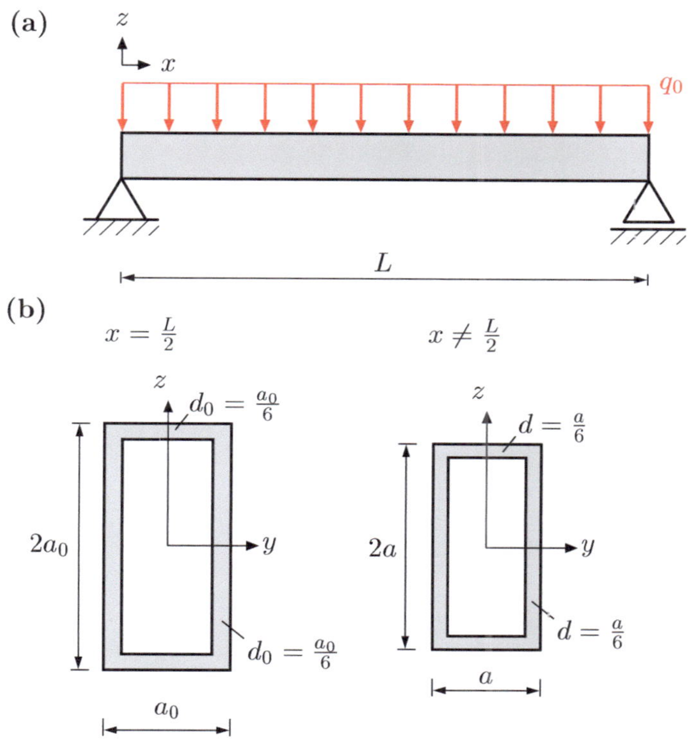

Fig. 2.21 Optimization of a simply supported beam under constant distributed load: **a** General configuration and **b** box-like cross sections

(d) The specific energy absorption SEA_0 of the cross section with constant values $(2a_0, a_0, d_0)$.
(e) The specific energy absorption SEA of the optimized cross section $(2a(x), a(x), d(x))$.
(f) What can be said about the technical realization of the contour $a = a(x)$?

Remarks:
$\int_0^L (-(x/L)^2 + x/L)^2 \mathrm{d}x \approx L/30$
$\int_0^L (-(x/L)^2 + x/L)^{2/3} \mathrm{d}x \approx 0.293342L.$

2.3.28 Optimization of a beam with I-profile along the longitudinal axis under the influence of a constant distributed load

For a simply supported beam (bending stiffness $E \times I(x)$, length L, see Fig. 2.22a) which is loaded by a constant distributed load q_0, optimize the cross section along the beam axis x taking into account a stress criterion (0.2-% strain limit $R_{p0\,2}$), see Fig. 2.22b. The cross-sectional dimensions in the middle at $x = L/2$ are given by the outer dimensions $\alpha a_0 \times a_0$ and the constant thickness $w_0 = a_0/6$. The cross-sectional ratios at any point $x \neq L/2$ are given by the external dimensions $\alpha a(x) \times$

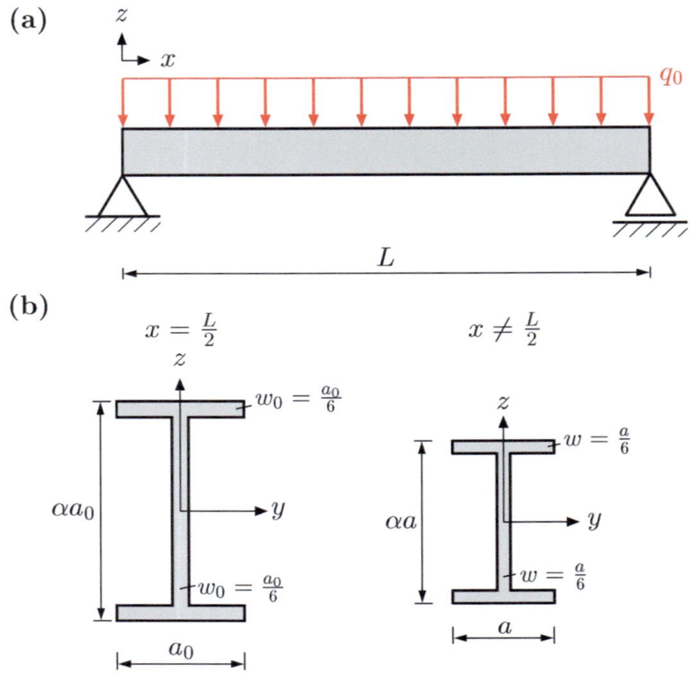

Fig. 2.22 Optimization of a simply supported beam under constant distributed load: **a** General configuration and **b** I cross sections

$a(x)$ and the thickness $w(x) = a(x)/6$, see Fig. 2.22b. *Consider the comments on integration at the end of the problem!*

Determine for $\alpha = 2$ and alternatively for $\alpha = 3$:

(a) The optimized beam contour $a = a(x)$ along the beam axis x (including sketch).
(b) An approximate value for the lightweight index M_0 of the cross section with constant values $(\alpha a_0, a_0, w_0)$.
(c) An approximate value for the lightweight index M of the optimized cross section $(\alpha a(x), a(x), w(x))$.
(d) The specific energy absorption SEA_0 of the cross section with constant values $(\alpha a_0, a_0, w_0)$.
(e) The specific energy absorption SEA of the optimized cross section with variable values $(\alpha a(x), a(x), w(x))$.
(f) What can be said about the technical realization of the contour $a = a(x)$?

Remarks:
$\int_0^L (-(x/L)^2 + x/L)^2 \mathrm{d}x \approx L/30$
$\int_0^L (-(x/L)^2 + x/L)^{2/3} \mathrm{d}x \approx 0.293342L.$

2.3.29 Optimization of a beam with I-profile along the longitudinal axis and the height-to-width ratio under the influence of a constant distributed load
For problem 2.3.28, determine the optimal α to maximize the specific energy absorption SEA of the optimized cross section.

2.3.30 Optimization of a beam along the longitudinal axis under the influence of two different single forces
For a cantilever beam (bending stiffness $E \times I(x)$, length L) which is loaded by two different single forces, optimize the box profile along the beam axis x taking into account a stress criterion (0.2-% strain limit $R_{p0.2}$), see Fig. 2.23a.

The cross-sectional dimensions at the fixed support at $x = 0$ (see Fig. 2.23b) are given by the outer diameter d_0 and the wall thickness $s_0 = d_0/5$. The cross-sectional ratios at any point $x > 0$ are shown in Fig. 2.23b. Determine:

(a) the beam contour $d = d(x)$ along the beam axis (including sketch),
(b) the specific energy absorption SEA of the optimized cross section.

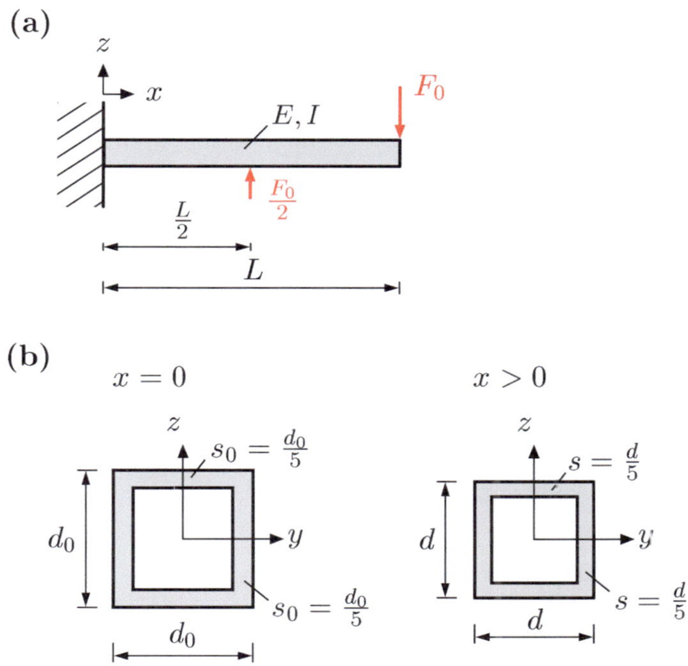

Fig. 2.23 Cantilever beam loaded by two different point loads: **a** General configuration and **b** cross sections

2.3.31 Optimization of a beam along the longitudinal axis under the influence of a single force and a single moment

For a cantilever beam (bending stiffness $E \times I(x)$, length L) loaded by a single force F_0 and a single moment $M_0 = \frac{F_0 L}{6}$, optimize the I-profile along the beam axis x taking into account a stress criterion (0.2-% strain limit $R_{p0.2}$), see Fig. 2.24a. The cross-sectional dimensions at the fixed support at $x = 0$ (see Fig. 2.24b) are given by the external dimensions a_0 and the wall thicknesses $w_0 = \frac{a_0}{6}$. The cross-sectional ratios at any point $x > 0$ are shown in Fig. 2.24b. Determine:

(a) the beam contour $a = a(x)$ along the beam axis (including sketch),
(b) the specific energy absorption SEA of the optimized cross section,
(c) compare the specific energy absorption SEA of the optimized cross section with the specific energy absorption for a beam with a constant cross section (a_0 and w_0).

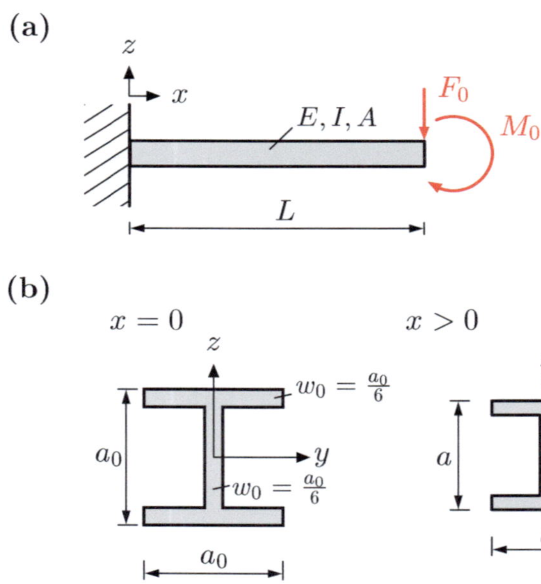

Fig. 2.24 Cantilever beam under the influence of a single force and a single moment: **a** General configuration and **b** I cross sections

2.3.32 Optimization of a cantilever beam along the longitudinal axis under the influence of a constant distributed load

For a cantilever beam (bending stiffness $E \times I(x)$, length L) subjected to a constant distributed load (q_0), optimize the box section along the beam axis x taking into account a stress criterion (0.2-% strain limit $R_{p0.2}$), see Fig. 2.25a. The cross-sectional dimensions at the fixed support $x = 0$ (see Fig. 2.25b) are given by the external dimension a_0 and the wall thickness $d_0 = \frac{a_0}{6}$. The cross-sectional ratios at any point $x > 0$ are shown in Fig. 2.25b.

First consider a beam with a constant cross section (see Fig. 2.25b (left)) and calculate:

(a) the cross-sectional area A_0, the axial second moment of area I_0 and the mass m_0,
(b) the specific energy absorption SEA_0 taking into account a stress criterion (0.2-% strain limit $R_{p0.2}$).

The optimized beam will now be considered. To do this, determine:

(c) the optimized beam dimension $a(x)$,
(d) the cross-sectional area $A(x)$, the axial second moment of area $I(x)$ and the mass m,
(e) the specific energy absorption SEA taking into account a stress criterion (0.2-% strain limit $R_{p0.2}$).

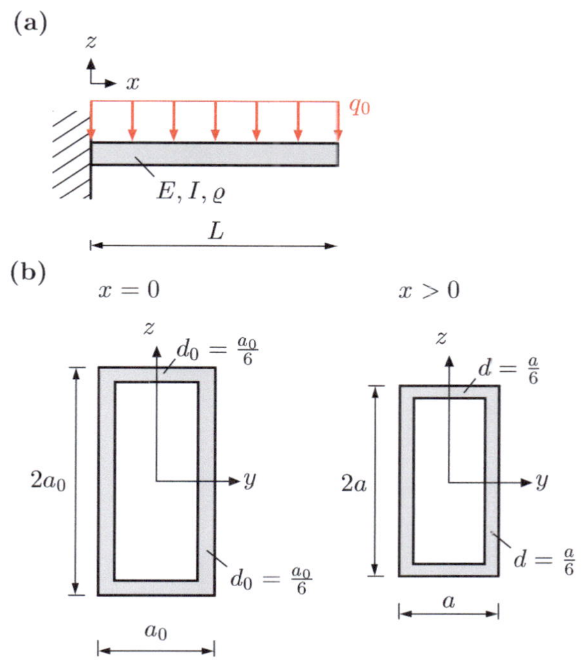

Fig. 2.25 Cantilever beam under the influence of a constant distributed load: **a** General configuration and **b** box-like cross sections

Finally, calculate the ratios m/m_0 and SEA/SEA_0. What would an external load have to look like so that the 'optimized' beam has the same cross section at every point x?

2.3.33 Optimization of a simply supported beam under constant distributed load: box cross section

For a simply supported beam (bending stiffness $E \times I(x)$, length L, see Fig. 2.26a) which is loaded by a constant distributed load q_0, optimize the rectangular cross section along the beam axis x taking into account a stress criterion (0.2-% strain limit $R_{p0.2}$), see Fig. 2.26b. The cross-sectional dimensions in the middle at $x = L/2$ are given by the height a_0, the width $b_0 = a_0/2$ and the constant thickness $d_0 = a_0/6$, and the cross-sectional ratios at any point $x \neq L/2$ are given by the height $a(x)$, the width $b(x) = a(x)/2$ and the thickness $d(x) = a(x)/6$, see Fig. 2.26b.

Determine the optimized beam contour $a = a(x)$ along the beam axis x (including sketch).

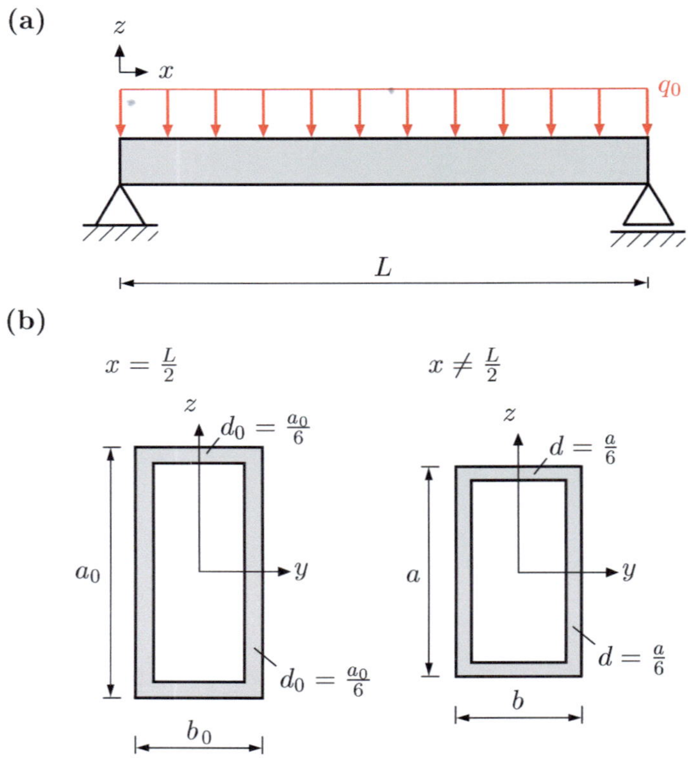

Fig. 2.26 Optimization of a simply supported beam under constant distributed load: **a** General configuration and **b** box-like cross sections

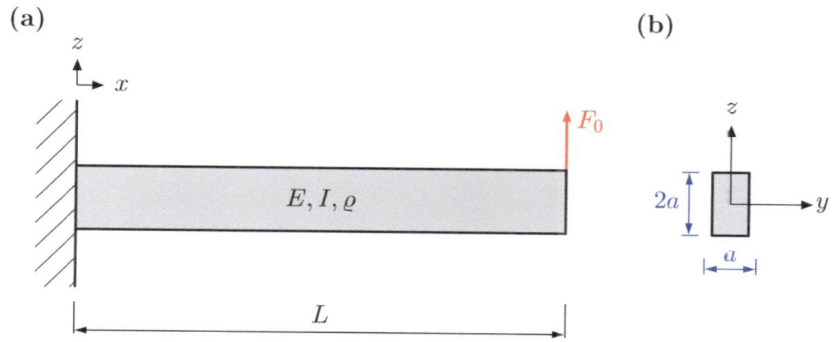

Fig. 2.27 **a** General configuration of the beam problem; **b** beam cross section

2.3.34 Optimization of a simply supported bending under constant distributed load: box cross section. Numerical calculation of the specific energy absorption
For problem 2.3.33 determine the specific energy absorption taking into account a stress criterion (0.2-% strain limit $R_{p0.2}$). The numerical calculation according to appendix A.1.4 is to be carried out for the reference cross section and the optimized cross section.

2.3.35 Optimal design of a cantilever beam (classical optimization problem)
Consider a cantilever beam fixed at one end as shown in Fig. 2.27. The beam is loaded by a single force F_0 and has constant material (E, ϱ) and geometric (I) properties along its longitudinal axis. The material is isotropic and homogeneous and the beam theory for thin beams (Euler-Bernoulli) can be used for this example.

Given are:

- Geometric dimension: $L = 2540$ mm.
- Material properties of the beam: Elastic modulus $E = 68948$ MPa, mass density $\varrho = 2691$ kg/m^3, 0.2-% strain limit $R_{p0.2} = 247$ MPa.
- Load: $F_0 = 2667$ N.

Determine the optimized cross-sectional dimension a under the condition that the effective stress does not exceed the initial yield stress. Furthermore, a maximum deflection of $u_z(L) = r_1 L$ with $r_1 = 0.03$ should not be exceeded. Based on a graphical representation of the objective function and the corresponding inequality constraints, the optimum is to be determined analytically.

2.3.36 Optimal design of a compression strut (classical optimization problem)
A compression strut is considered according to Fig. 2.28. The strut is loaded by a single force F_0 and has constant material (E, ϱ) and geometric (I, A) properties along its longitudinal axis. Furthermore, the material can be considered isotropic and homogeneous.

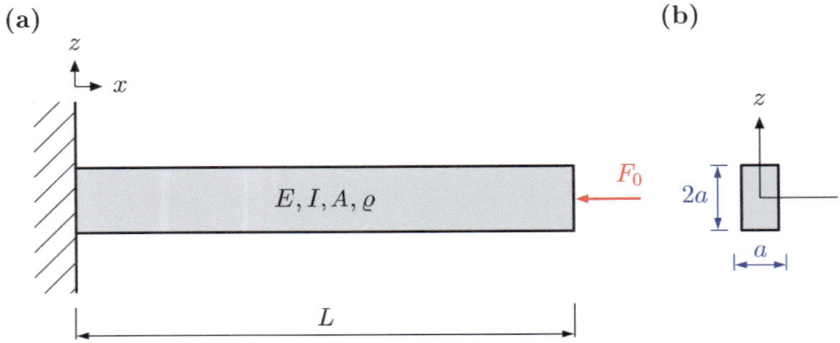

Fig. 2.28 **a** General configuration of the compression strut problem; **b** cross section

Given are:

- Geometric dimension: $L = 1000\,\text{mm}$.
- Material properties of the strut: Elastic modulus $E = 70000\,\text{MPa}$, mass density $\varrho = 2691\,\text{kg/m}^3$, 0.2-% strain limit $R_{\text{p}0.2} = 247\,\text{MPa}$.
- Load: $F_0 = 2667\,\text{N}$.

Determine the optimized cross-sectional dimension a under the condition that the effective stress does not exceed the initial yield stress. Furthermore, no buckling should occur in the elastic region. The optimum is to be determined analytically based on a graphical representation of the objective function and the corresponding inequality constraints.

2.3.37 Optimal design of a short cantilever beam (classical optimization problem)

Consider a short cantilever beam as shown in Fig. 2.29. The beam is loaded by a single force F_0 and has constant material (E, ϱ) and geometric (I) properties along its longitudinal axis. The material is isotropic and homogeneous and the beam theory for thick beams (i.e., consideration of the acting shear stress) can be used for this example.

Fig. 2.29 **a** General configuration of the short cantilever beam problem; **b** cross section

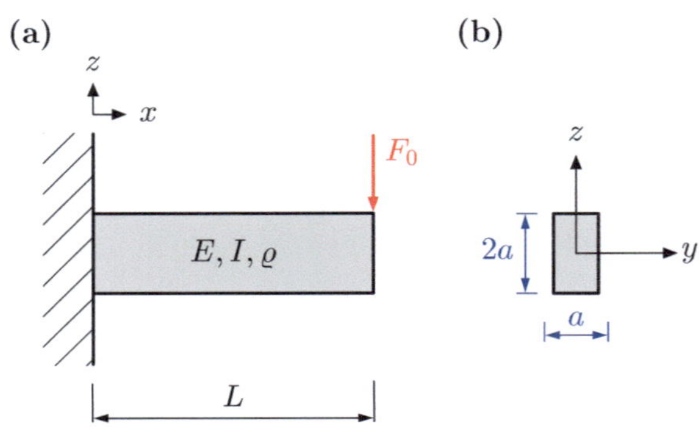

Given are:

- Geometric dimension: $L = 846.33\,\text{mm}$.
- Material properties of the beam: elastic modulus $E = 68948\,\text{MPa}$, mass density $\varrho = 2691\,\text{kg/m}^3$, 0.2-% strain limit $R_{\text{p0.2}} = 247\,\text{MPa}$.
- Load: $F_0 = 2667\,\text{N}$.

Determine the optimized cross-sectional dimension a under the condition that the acting normal and shear stresses do not exceed the corresponding initial yield stress. The initial shear yield stress can be approximated using the Tresca hypothesis. The normal stress distribution can be assumed to be linear and the shear stress distribution to be parabolic over the beam height. Based on a graphical representation of the objective function and the corresponding inequality restrictions, the optimum is to be determined analytically.

2.3.38 Optimal design of a cantilever beam: constant rectangular cross section (classical optimization problem)

Consider a cantilever beam clamped at one end as shown in Fig. 2.30. The beam is loaded by a single force F_0 and has constant material (E, ϱ) and geometric (I) properties along its longitudinal axis. The material is isotropic and homogeneous and the beam theory for thin beams (Euler-Bernoulli) can be used for this problem.

Given are:

- Geometric dimension: $L = 2540\,\text{mm}$.
- Material properties of the beam: elastic modulus $E = 68948\,\text{MPa}$, mass density $\varrho = 2691\,\text{kg/m}^3$, 0.2-% strain limit $R_{\text{p0.2}} = 247\,\text{MPa}$.
- Load: $F_0 = 2667\,\text{N}$.

Determine the optimized cross-sectional dimensions a and b under the condition that the acting normal stress does not exceed the initial yield stress. Furthermore, a maximum deflection of $u_z(L) = r_1 L$ with $r_1 = 0.03$ should not be exceeded. Finally,

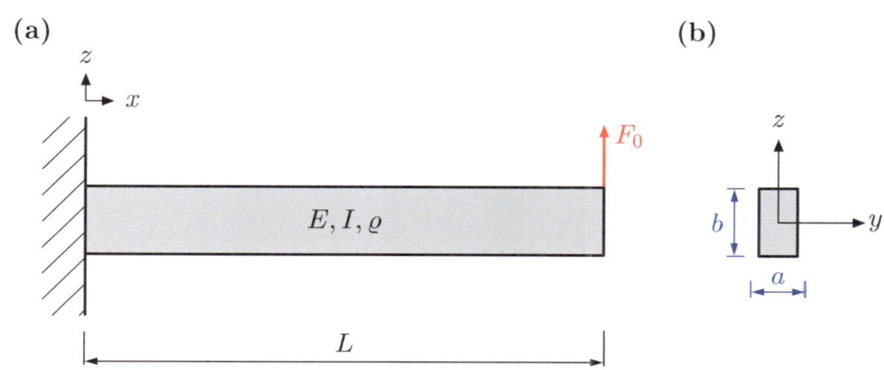

Fig. 2.30 General configuration of the beam problem; **a** General configuration of the beam problem; **b** beam cross section

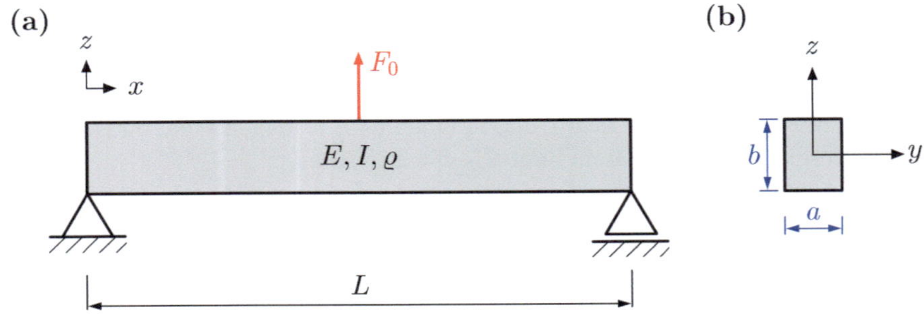

Fig. 2.31 General configuration of a simply supported beam under 3-point bending; **a** General configuration of a simply supported beam under 3-point bending; **b** cross section

the height-to-width ratio should be limited to $b \leq 20a$ in order to avoid instabilities. Based on a graphical representation of the objective function and the corresponding inequality constraints, the optimum is to be determined analytically. Compare the optimized geometry with the results from Exercise 2.3.35.

2.3.39 Optimal design of a simply supported beam: constant rectangular cross section (classical optimization problem)

Consider a simply supported beam as shown in Fig. 2.31. The beam is loaded in the middle by a single force F_0 and has constant material (E, ϱ) and geometric (I) properties along its longitudinal axis. The material is isotropic and homogeneous and the beam theory for thin beams (Euler-Bernoulli) can be used for this problem. Given are:

- Geometric dimension: $L = 2540\,\text{mm}$.
- Material properties of the beam: elastic modulus $E = 68948\,\text{MPa}$, mass density $\varrho = 2691\,\text{kg/m}^3$, 0.2-% strain limit $R_{\text{p}0.2} = 247\,\text{MPa}$, shear yield stress $\tau_{\text{p}} = R_{\text{p}0.2}/2$.
- Load: $F_0 = 2667\,\text{N}$.

Determine the optimized cross-sectional dimensions a and b under the condition that the acting normal and shear stresses do not exceed the corresponding initial yield stresses. Furthermore, a maximum deflection of $u_z(L) = r_1 L$ with $r_1 = 0.03$ or $r_1' = 0.003$ should not be exceeded. Finally, the height-to-width ratio (aspect ratio) should be limited to $b \leq 20a$ in order to avoid instabilities. Based on a graphical representation of the objective function and the corresponding inequality constraints, the optimum is to be determined analytically.

2.3.40 Optimal design of a compression strut: constant rectangular cross section (classical optimization problem)

A compression strut is considered according to Fig. 2.32. The strut is loaded by a single force F_0 and has constant material (E, ϱ) and geometric (I, A) properties

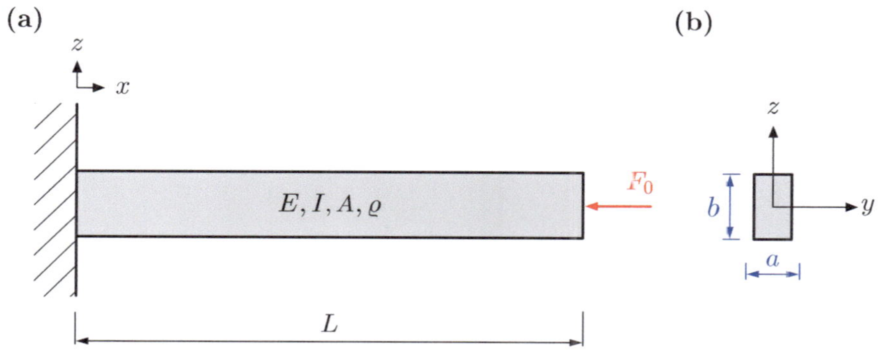

Fig. 2.32 **a** General configuration of the compression strut problem; **b** cross section

along its longitudinal axis. Furthermore, the material can be considered isotropic and homogeneous.
Given are:

- Geometric dimension: $L = 1000$ mm.
- Material properties of the strut: elastic modulus $E = 70000$ MPa, mass density $\varrho = 2691\,\text{kg/m}^3$, 0.2-% strain limit $R_{\text{p}0.2} = 247$ MPa.
- Load: $F_0 = 2667$ N.

Determine the optimized cross-sectional dimensions a and b under the condition that the effective stress does not exceed the initial yield stress. Furthermore, no buckling should occur in the elastic range. Based on a graphical representation of the objective function and the corresponding inequality constraints, the optimum is to be determined analytically. Compare the optimized geometry with the results from Exercise 2.3.36.

2.3.41 Optimal design of a short cantilever beam: constant rectangular cross section (classical optimization problem)

Consider a short, cantilevered bending beam as shown in Fig. 2.33. The beam is loaded by a single force F_0 and has constant material (E, ϱ) and geometric (I) properties along its longitudinal axis. The material is isotropic and homogeneous and the beam theory for thick beams (i.e., consideration of the acting shear stress) can be used for this example.
Given are:

- Geometric dimension: $L = 846.33$ mm.
- Material properties of the beam: elastic modulus $E = 68948$ MPa, mass density $\varrho = 2691\,\text{kg/m}^3$, 0.2-% strain limit $R_{\text{p}0.2} = 247$ MPa.
- Load: $F_0 = 2667$ N.

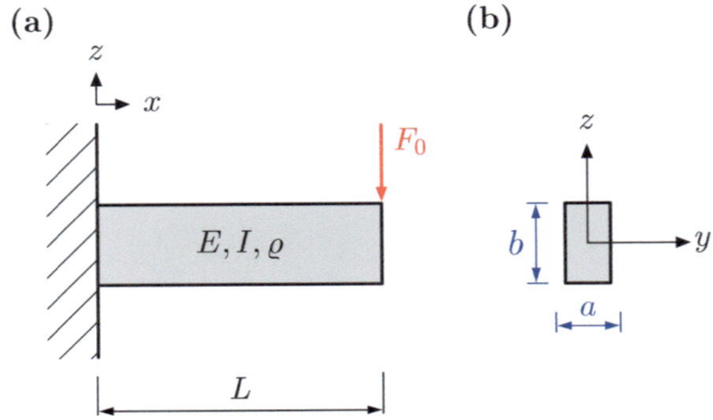

Fig. 2.33 General configuration of the short cantilever beam problem; **a** General configuration of the short cantilever beam problem; **b** cross section

Determine the optimized cross-sectional dimensions a and b under the condition that the acting normal and shear stresses do not exceed the corresponding initial yield stresses. The initial shear yield stress can be approximated using the Tresca hypothesis. The normal stress distribution can be assumed to be linear and the shear stress distribution to be parabolic over the beam height. Furthermore, the height-to-width ratio should be limited to $b \leq 20a$ in order to avoid instabilities. Based on a graphical representation of the objective function and the corresponding inequality constraints, the optimum is to be determined analytically. Compare the optimized geometry with the results from Exercise 2.3.37.

2.3.42 Optimal design of a stepped cantilever beam with two sections (classical optimization problem)

Consider a cantilever beam that is fixed on one side as shown in Fig. 2.34 and loaded at the right-hand end by a single force F_0 in the negative Z-direction. The beam is divided into two sections of length $L_\text{I} = L_\text{II} = L/2$. The material of each section is the same, i.e. $E_\text{I} = E_\text{II} = E$, but the cross sections are different in both sections. Optimize the two cross sections, i.e. the heights b_i ($i = \text{I}, \text{II}$), whereby the width of the two sections (a) remains identical.
Given are:

- Geometric dimensions: $L = 2540\,\text{mm}$, $a = 45.16\,\text{mm}$.
- Material properties of the beam: elastic modulus $E = 68948\,\text{MPa}$, mass density $\varrho = 2691\,\text{kg/m}^3$, 0.2-% strain limit $R_{\text{p0.2}} = 247\,\text{MPa}$.
- Load: $F_0 = 2667\,\text{N}$.

Determine the optimized cross-sectional dimensions b_i under the condition that the acting normal and shear stresses do not exceed the corresponding initial yield stresses. The initial shear yield stress can be approximated using the Tresca hypothesis. The normal stress distribution can be assumed to be linear and the shear stress distribution to be parabolic over the beam height. Furthermore, a maximum deflection of $|u_z(L)| = r_1 L$ with $r_1 = 0.06$ should not be exceeded. Finally, the height-to-width ratio should be limited to $b \leq 20a$ to avoid instabilities. Based on a graphical representation of the objective function and the corresponding inequality constraints, the

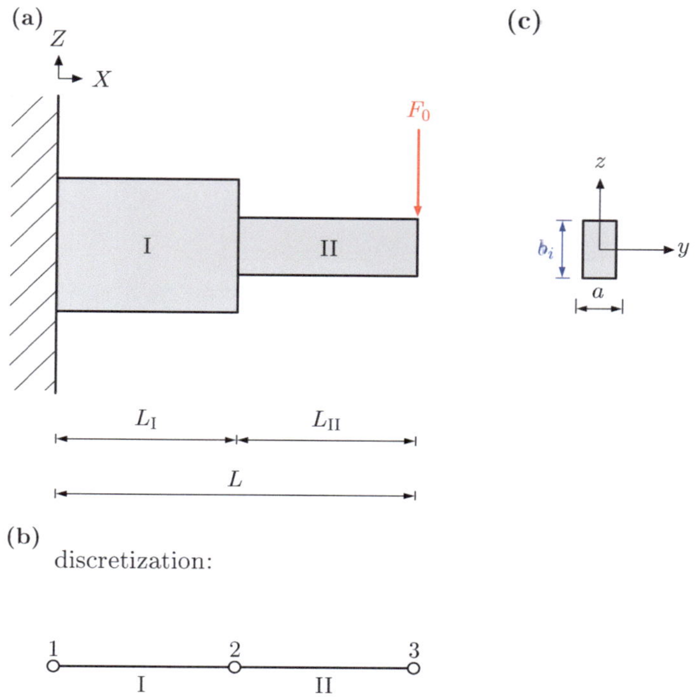

Fig. 2.34 Cantilever stepped beam with two sections: **a** General configuration, **b** discretization, and **c** cross section of element i

optimum is to be determined analytically. Use a finite element approach ('hand calculation') to determine the deformations and internal reactions, see Öchsner (2016, 2018); Javanbakht und Öchsner (2018).

2.3.43 Optimal design of a simply supported beam with three sections (classical optimization problem)

Consider a simply supported beam as shown in Fig. 2.35, which is loaded in the middle by a single force F_0 in the negative Z-direction. The beam is divided into three sections of length $L_{\mathrm{I}} = L_{\mathrm{II}} = L_{\mathrm{III}} = L/3$. The material of each section is the same, i.e. $E_{\mathrm{I}} = E_{\mathrm{II}} = E_{\mathrm{III}} = E$, but the middle cross section has different dimensions than the two outer sections (I = III). Optimize the cross sections, i.e. the heights b_i ($i = \mathrm{I, II, III}$), whereby the width of the three sections (a) remains identical. For simplification, consider the symmetry of the problem.
Given are:

- Geometric dimensions: $L = 2540\,\text{mm}$, $a = 45.16\,\text{mm}$.
- Material properties of the beam: elastic modulus $E = 68948\,\text{MPa}$, mass density $\varrho = 2691\,\text{kg/m}^3$, 0.2-% strain limit $R_{\text{p0.2}} = 247\,\text{MPa}$.
- Load: $F_0 = 2667\,\text{N}$.

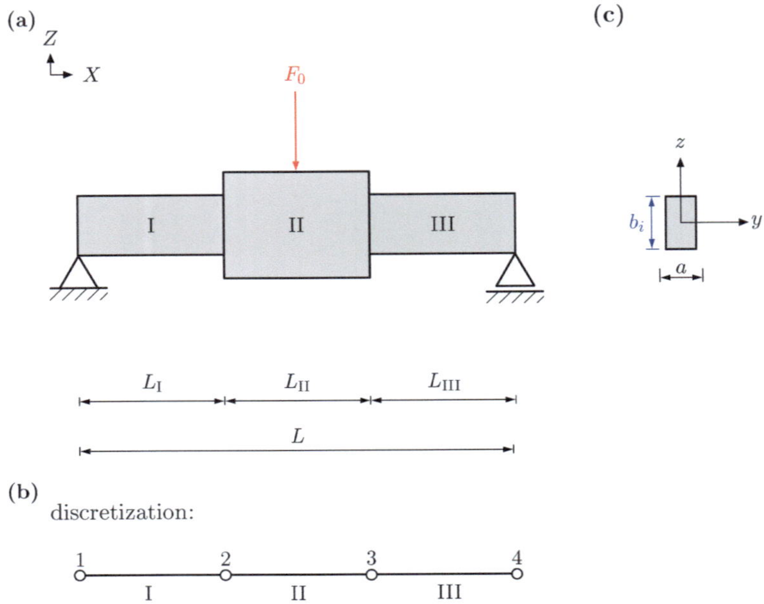

Fig. 2.35 Simply supported beam with three sections: **a** General configuration, **b** discretization, and **c** cross section of element i

Determine the optimized cross-sectional dimensions b_i under the condition that the acting normal and shear stresses do not exceed the corresponding initial yield stresses. The initial shear yield stress can be approximated using the Tresca hypothesis. The normal stress distribution can be assumed to be linear and the shear stress distribution to be parabolic over the beam height. Furthermore, a maximum deflection of $|u_z(L/2)| = r_1 L$ with $r_1 = 0.01$ should not be exceeded. Finally, the height-to-width ratio should be limited to $b \leq 20a$ to avoid instabilities. Based on a graphical representation of the objective function and the corresponding inequality constraints, the optimum is to be determined analytically. Use a finite element approach ('hand calculation') to determine the deformations and internal reactions, see Öchsner (2016, 2018); Javanbakht und Öchsner (2018).

2.4 Sandwich Structures: Basic Mechanical Load Cases

2.4.44 Average bending stiffness of a sandwich

Determine the general average bending stiffness $\overline{EI}_y$ for the sandwich shown in Fig. 2.36, which consists of three layers and a cross-section as an I-profile. Afterwards, simplify the result for the special case $b^{\mathrm{C}} = b^{\mathrm{F}}/3$ and $\Delta h^{\mathrm{F}} = \Delta h^{\mathrm{C}}/4$.

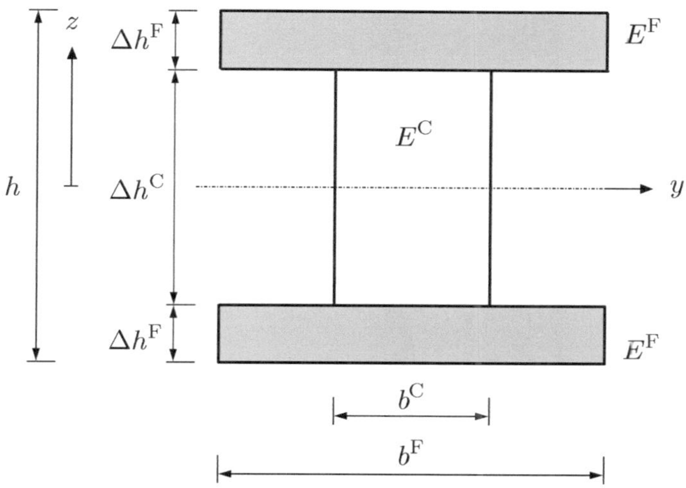

Fig. 2.36 Sandwich of three layers and cross section as I-profile

2.4.45 Calculation of the shear stress distribution for a circular cross section

A beam with a circular cross section (radius R) is given. Calculate the shear stress distribution τ_{zx} over the cross section under the influence of a shear force $Q_z(x)$. Consider the distribution in the center of the circular cross section, i.e. for $y = 0$.

2.4.46 Limit case of the normalized stiffness of a sandwich

Simplify the expression for the normalized stiffness of a sandwich according to Eq. $(5.57)_{\mathrm{TB}}$, i.e.,

$$\frac{\overline{EI_y}}{\overline{EI_{y,\mathrm{ref}}}} = \left[\frac{1}{4} + \frac{3}{4}\left(\frac{\Delta h^{\mathrm{C}}}{\Delta h^{\mathrm{F}}} + 1\right)^2 + \frac{1}{8}\left(\frac{E^{\mathrm{C}}}{E^{\mathrm{F}}}\right)\left(\frac{\Delta h^{\mathrm{C}}}{\Delta h^{\mathrm{F}}}\right)^3\right]. \tag{2.5}$$

for the limiting case $E^{\mathrm{F}} = E^{\mathrm{C}} = E$ and $\Delta h^{\mathrm{F}} = \Delta h^{\mathrm{C}} = \frac{1}{3}$. This is a homogeneous body of total height '1'. Clarify how the result to understand (why is the ratio not equal to '1'?).

2.4.47 Deflection of a sandwich beam with distributed load using the partial deflection method

For the sandwich beam shown in Fig. 2.37, the bending line $u_z(x)$ and the maximum value of the deflection must be determined. It should be assumed that it is a beam with thin face sheets and a soft core. Furthermore, calculate the general ratio of the partial deflections and also provide the numerical value for $L = 2000$ mm, $\Delta h^{\mathrm{C}} = 150$ mm, $\Delta h^{\mathrm{F}} = 5$ mm, $E^{\mathrm{F}} = 74000$ MPa and $G^{\mathrm{C}} = 11$ MPa.

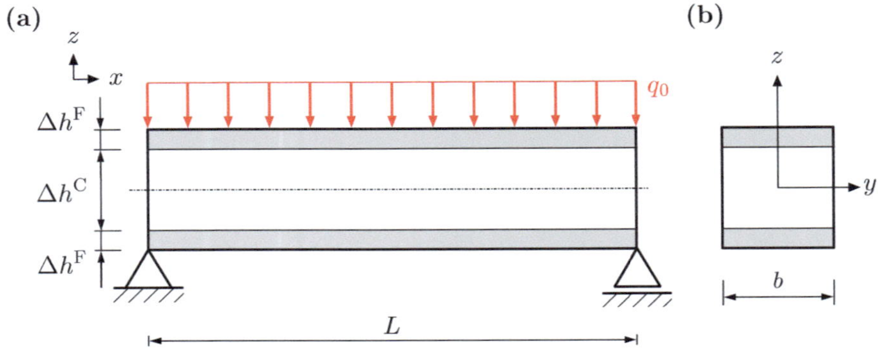

Fig. 2.37 Sandwich beam with constant distributed load: **a** boundary conditions and external load; **b** beam cross section

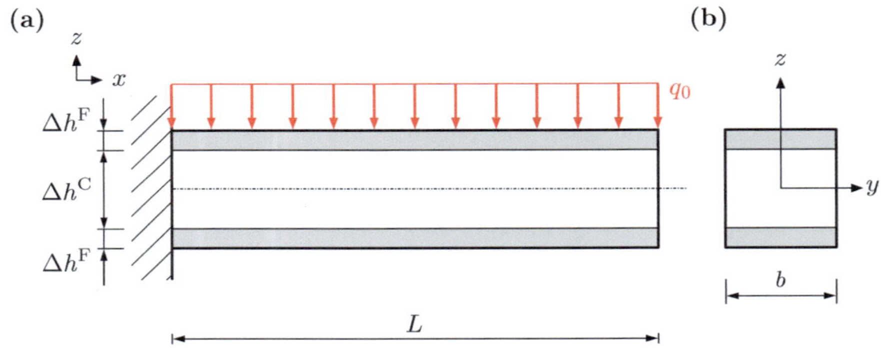

Fig. 2.38 Sandwich beam with constant distributed load: **a** boundary condition and external load; **b** beam cross section

2.4.48 Deflection of a sandwich beam (cantilever) with distributed load using the partial deflection method

For the sandwich beam shown in Fig. 2.38, the bending line $u_z(x)$ and the maximum value of the deflection must be determined. It should be assumed that it is a beam with thin face sheets and a soft core. Furthermore, calculate the general ratio of the partial deflections and also provide the numerical value for $L = 2000\,\text{mm}$, $\Delta h^{\mathrm{C}} = 150\,\text{mm}$, $\Delta h^{\mathrm{F}} = 5\,\text{mm}$, $E^{\mathrm{F}} = 74000\,\text{MPa}$ and $G^{\mathrm{C}} = 11\,\text{MPa}$.

2.4.49 Deflection of a sandwich beam (cantilever) with shear force using the partial deflection method

For the sandwich beam shown in Fig. 2.39, the bending line $u_z(x)$ and the maximum deflection value must be determined. It should be assumed that this is a beam with thin face sheets and a soft core. Furthermore, calculate the general ratio of the partial deflections and also provide the numerical value for $L = 2000\,\text{mm}$, $\Delta h^{\mathrm{C}} = 150\,\text{mm}$, $\Delta h^{\mathrm{F}} = 5\,\text{mm}$, $E^{\mathrm{F}} = 74000\,\text{MPa}$ and $G^{\mathrm{C}} = 11\,\text{MPa}$.

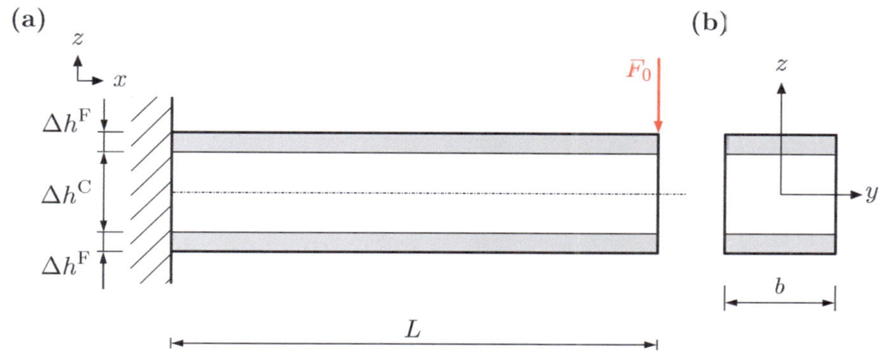

Fig. 2.39 Sandwich beam with shear force: **a** boundary condition and external load; **b** beam cross section

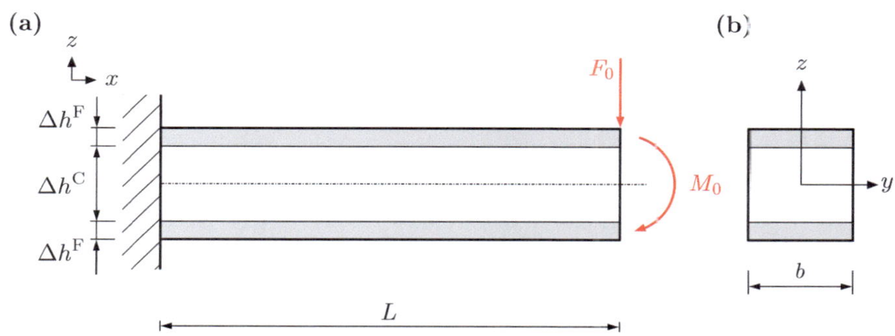

Fig. 2.40 Sandwich beam with shear force and single moment: **a** boundary condition and external loads; **b** beam cross section

2.4.50 Deflection of a sandwich beam (cantilever) with shear force and bending moment using the partial deflection method

For the sandwich beam shown in Fig. 2.40, the bending line $u_z(x)$ and the maximum deflection value must be determined. In addition to the single force F_0, a single moment of magnitude $M_0 = \frac{1}{2} F_0 L$ acts on the beam. It should be assumed that this is a beam with thin face sheets and a soft core. Furthermore, calculate the general ratio of the partial deflections and also provide the numerical value for $L = 2000$ mm, $\Delta h^C = 150$ mm, $\Delta h^F = 5$ mm, $E^F = 74000$ MPa and $G^C = 11$ MPa.

2.5 Sandwich Structures: Limit Load

2.5.51 Failure analysis of a sandwich beam with distributed load

Carry out the corresponding proofs of strength for the sandwich beam shown in Fig. 2.41. The sandwich consists of a hard foam core material bonded to two aluminum face sheets. The thickness of the adhesive layer can be neglected in the

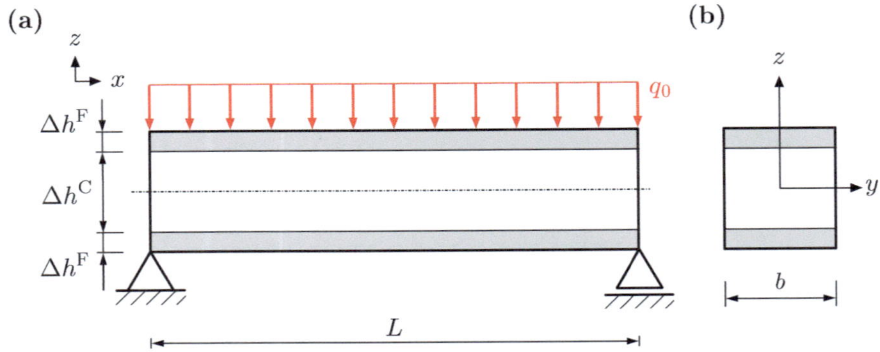

Fig. 2.41 Sandwich beam with constant distributed load: **a** boundary conditions and external load; **b** beam cross section

calculation. Furthermore, it should be assumed that the value of the distributed load q_0 has already been multiplied by a sufficient safety factor.
Given are:

- Geometric dimensions: $L = 2000\,\text{mm}$, $b = 200\,\text{mm}$, $\Delta h^{\text{C}} = 150\,\text{mm}$, $\Delta h^{\text{F}} = 5\,\text{mm}$.
- Material properties of the hard foam core: $E^{\text{C}} = 30\,\text{MPa}$, $\nu^{\text{C}} = 0.364$, $\tau^{\text{C}}_{\text{aB}} = 0.5\,\text{MPa}$, $R^{\text{C}}_{\text{m}} = 0.90\,\text{MPa}$, $\sigma^{\text{C}}_{\text{dB}} = 0.38\,\text{MPa}$.
- Material properties of the face sheets: $E^{\text{F}} = 74000\,\text{MPa}$, $R^{\text{F}}_{\text{p0.2}} = 364\,\text{MPa}$.
- Material property of the adhesive: $\tau_{\text{aB}} = 37.1\,\text{MPa}$.
- External load: $q_0 = 2.5\,\frac{\text{N}}{\text{mm}}$.

2.5.52 Failure analysis of a sandwich beam (cantilever) with distributed load

Carry out the corresponding proofs of strength for the sandwich beam shown in Fig. 2.42. The sandwich consists of a hard foam core material bonded to two aluminum face sheets. The thickness of the adhesive layer can be neglected in the calculation. Furthermore, it should be assumed that the value of the distributed load q_0 has already been multiplied by a sufficient safety factor.
Given are:

- Geometric dimensions: $L = 2000\,\text{mm}$, $b = 200\,\text{mm}$, $\Delta h^{\text{C}} = 150\,\text{mm}$, $\Delta h^{\text{F}} = 5\,\text{mm}$.
- Material properties of the hard foam core: $E^{\text{C}} = 30\,\text{MPa}$, $\nu^{\text{C}} = 0.364$, $\tau^{\text{C}}_{\text{aB}} = 0.5\,\text{MPa}$, $R^{\text{C}}_{\text{m}} = 0.90\,\text{MPa}$, $\sigma^{\text{C}}_{\text{dB}} = 0.38\,\text{MPa}$.
- Material properties of the face sheets: $E^{\text{F}} = 74000\,\text{MPa}$, $R^{\text{F}}_{\text{p0.2}} = 364\,\text{MPa}$.
- Material property of the adhesive: $\tau_{\text{aB}} = 37.1\,\text{MPa}$.
- External load: $q_0 = 2.5\,\frac{\text{N}}{\text{mm}}$.

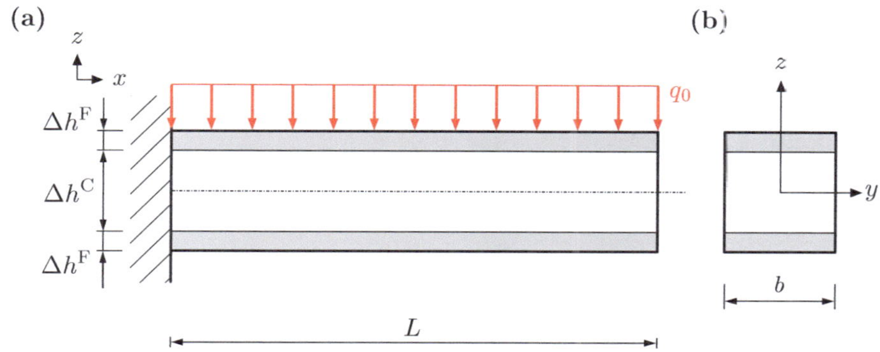

Fig. 2.42 Sandwich beam (cantilever) with constant distributed load: **a** boundary conditions and external load; **b** beam cross section

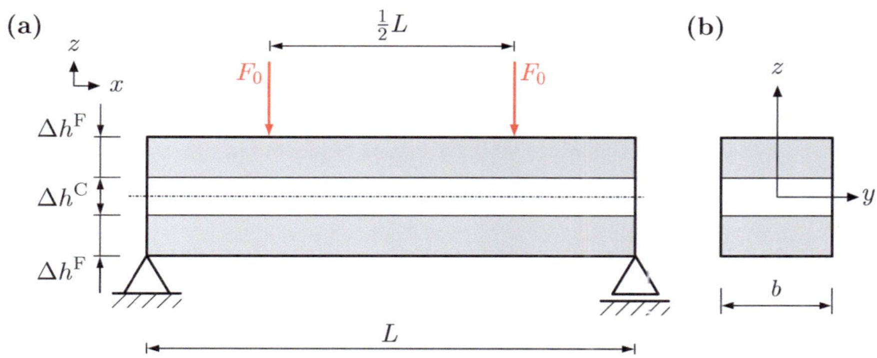

Fig. 2.43 Sandwich beam under 4-point bending: **a** boundary conditions and external loads; **b** beam cross section

2.5.53 Failure analysis of a sandwich beam under 4-point bending

Carry out the corresponding proofs of strength for the sandwich beam shown in Fig. 2.43. The sandwich is constructed from a synthetic core material bonded to two aluminum face sheets. The thickness of the adhesive layer can be neglected in the calculation. Furthermore, it should be assumed that the value of the force F_0 has already been multiplied by a sufficient safety factor.

Given are:

- Geometric dimensions: $L = 2000\,\mathrm{mm}$, $b = 200\,\mathrm{mm}$, $\Delta h^{\mathrm{C}} = 50\,\mathrm{mm}$, $\Delta h^{\mathrm{F}} = 50\,\mathrm{mm}$.
- Material properties of the synthetic core: $E^{\mathrm{C}} = 25000\,\mathrm{MPa}$, $\nu^{\mathrm{C}} = 0.4$, $\tau^{\mathrm{C}}_{\mathrm{aB}} = 40\,\mathrm{MPa}$, $R^{\mathrm{C}}_{\mathrm{m}} = 80\,\mathrm{MPa}$, $\sigma^{\mathrm{C}}_{\mathrm{dB}} = 60\,\mathrm{MPa}$.
- Material properties of the face sheets: $E^{\mathrm{F}} = 74000\,\mathrm{MPa}$, $R^{\mathrm{F}}_{\mathrm{p0.2}} = 364\,\mathrm{MPa}$.
- Material property of the adhesive: $\tau_{\mathrm{aB}} = 37.1\,\mathrm{MPa}$.
- External load: $F_0 = 2500\,\mathrm{N}$.

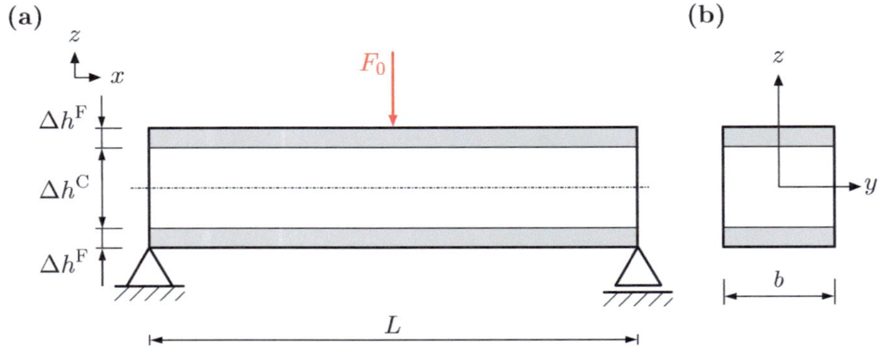

Fig. 2.44 Sandwich beam under 3-point bending: **a** boundary conditions and external load; **b** beam cross section

2.5.54 Failure analysis of a sandwich beam under 3-point bending with point load

Carry out the corresponding proofs of strength for the sandwich beam shown in Fig. 2.44. The sandwich consists of a plastic core bonded to two aluminum face sheets. The thickness of the adhesive layer can be neglected in the calculation. Furthermore, it should be assumed that the value of the force F_0 has already been multiplied by a sufficient safety factor.
Given are:

- Geometric dimensions: $L = 2000\,\mathrm{mm}$, $b = 200\,\mathrm{mm}$, $\Delta h^{\mathrm{C}} = 100\,\mathrm{mm}$, $\Delta h^{\mathrm{F}} = 25\,\mathrm{mm}$.
- Material properties of the plastic core: $E^{\mathrm{C}} = 1500\,\mathrm{MPa}$, $\nu^{\mathrm{C}} = 0.4$, $\tau^{\mathrm{C}}_{\mathrm{aB}} = 40\,\mathrm{MPa}$, $R^{\mathrm{C}}_{\mathrm{m}} = 80\,\mathrm{MPa}$, $\sigma^{\mathrm{C}}_{\mathrm{dB}} = 60\,\mathrm{MPa}$.
- Material properties of the face sheets: $E^{\mathrm{F}} = 74000\,\mathrm{MPa}$, $R^{\mathrm{F}}_{\mathrm{p0.2}} = 364\,\mathrm{MPa}$.
- Material property of the adhesive: $\tau_{\mathrm{aB}} = 37.1\,\mathrm{MPa}$.
- External load: $F_0 = 5000\,\mathrm{N}$.

2.5.55 Failure analysis of a sandwich beam under 3-point bending with distributed load

Carry out the corresponding proofs of strength for the sandwich beam shown in Fig. 2.45. The sandwich consists of a plastic core bonded to two aluminum face sheets. The thickness of the adhesive layer can be neglected in the calculation. Furthermore, it should be assumed that the value of the distributed load q_0 has already been multiplied by a sufficient safety factor.

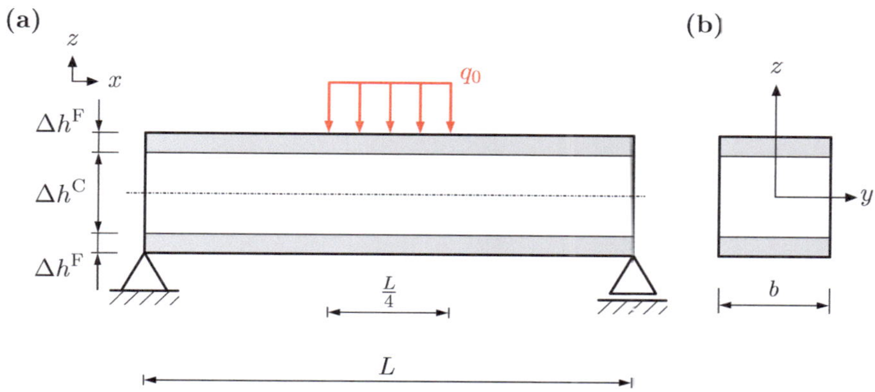

Fig. 2.45 Sandwich beam under 3-point bending with distributed load: **a** boundary conditions and external load; **b** beam cross section

Given are:

- Geometric dimensions: $L = 2000\,\text{mm}$, $b = 200\,\text{mm}$, $\Delta h^C = 100\,\text{mm}$, $\Delta h^F = 25\,\text{mm}$.
- Material properties of the plastic core: $E^C = 1500\,\text{MPa}$, $\nu^C = 0.4$, $\tau_{aB}^C = 40\,\text{MPa}$, $R_m^C = 80\,\text{MPa}$, $\sigma_{dB}^C = 60\,\text{MPa}$.
- Material properties of the face sheets: $E^F = 74000\,\text{MPa}$, $R_{p0.2}^F = 364\,\text{MPa}$.
- Material property of the adhesive: $\tau_{aB} = 37.1\,\text{MPa}$.
- External load: $q_0 = 10\,\text{N/mm}$.

2.5.56 Failure analysis of a sandwich beam under 5-point bending with point loads

Carry out the corresponding proofs of strength for the sandwich beam shown in Fig. 2.46. The sandwich consists of a plastic core bonded to two aluminum face sheets. The thickness of the adhesive layer can be neglected in the calculation. Furthermore, it should be assumed that the value of the forces F_0 is already multiplied by a sufficient safety factor.
Given are:

- Geometric dimensions: $L = 2000\,\text{mm}$, $b = 200\,\text{mm}$, $\Delta h^C = 100\,\text{mm}$, $\Delta h^F = 25\,\text{mm}$.
- Material properties of the plastic core: $E^C = 1500\,\text{MPa}$, $\nu^C = 0.4$, $\tau_{aB}^C = 40\,\text{MPa}$, $R_m^C = 80\,\text{MPa}$, $\sigma_{dB}^C = 60\,\text{MPa}$.
- Material properties of the face sheets: $E^F = 74000\,\text{MPa}$, $R_{p0.2}^F = 364\,\text{MPa}$.
- Material property of the adhesive: $\tau_{aB} = 37.1\,\text{MPa}$.
- External load: $F_0 = 2500\,\text{N}$.

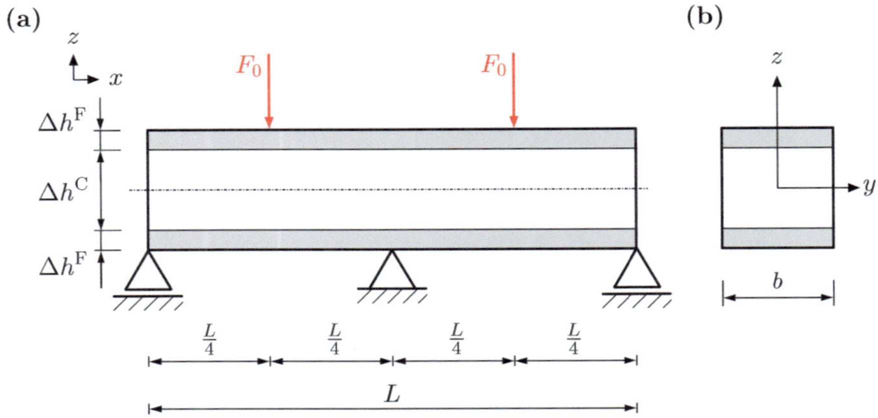

Fig. 2.46 Sandwich beam under 5-point bending with point loads: **a** boundary conditions and external loads; **b** beam cross section

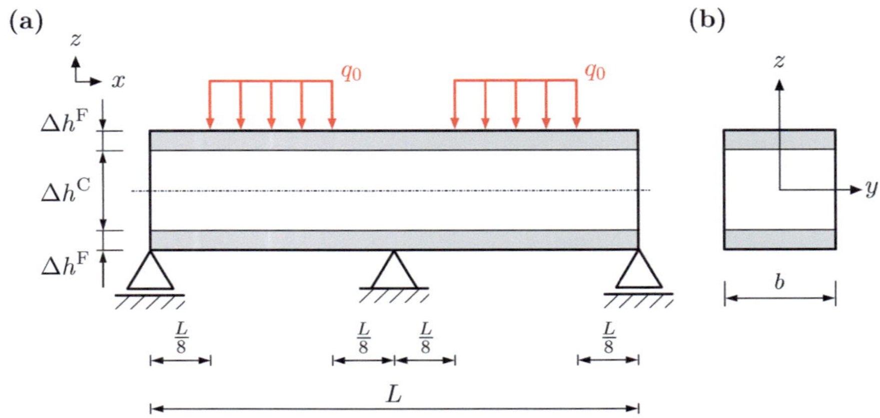

Fig. 2.47 Sandwich beam under 5-point bending with distributed loads: **a** boundary conditions and external loads; **b** beam cross section

2.5.57 Failure analysis of a sandwich beam under 5-point bending with distributed loads

Carry out the corresponding proofs of strength for the sandwich beam shown in Fig. 2.47. The sandwich consists of a plastic core bonded to two aluminum face sheets. The thickness of the adhesive layer can be neglected in the calculation. Furthermore, it should be assumed that the value of the distributed load q_0 has already been multiplied by a sufficient safety factor.

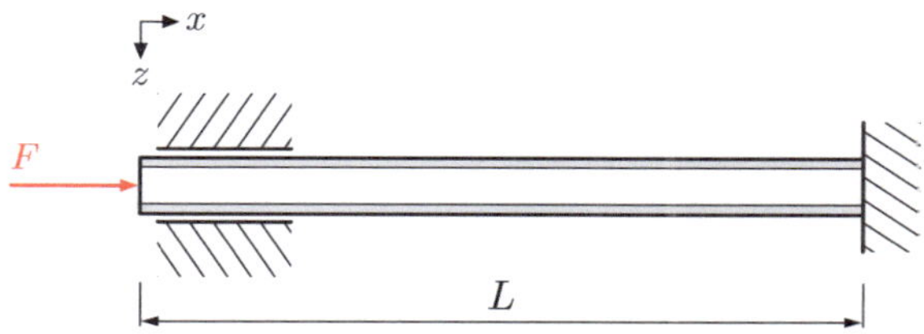

Fig. 2.48 Global instability failure of a sandwich beam clamped on both sides under compressive loading

Given are:

- Geometric dimensions: $L = 2000\,\text{mm}$, $b = 200\,\text{mm}$, $\Delta h^{\text{C}} = 100\,\text{mm}$, $\Delta h^{\text{F}} = 25\,\text{mm}$.
- Material properties of the plastic core: $E^{\text{C}} = 1500\,\text{MPa}$, $\nu^{\text{C}} = 0.4$, $\tau_{\text{aB}}^{\text{C}} = 40\,\text{MPa}$, $R_{\text{m}}^{\text{C}} = 80\,\text{MPa}$, $\sigma_{\text{dB}}^{\text{C}} = 60\,\text{MPa}$.
- Material properties of the face sheets: $E^{\text{F}} = 74000\,\text{MPa}$, $R_{\text{p0.2}}^{\text{F}} = 364\,\text{MPa}$.
- Material property of the adhesive: $\tau_{\text{aB}} = 37.1\,\text{MPa}$.
- External load: $q_0 = 5\,\text{N/mm}$.

2.5.58 Global instability failure of a sandwich beam clamped on both sides under compressive loading

Derive the buckling force for the sandwich beam shown in Fig. 2.48 assuming thin face sheets and a soft core. Compare the result with the classic solution according to Euler for the 4th support condition.

2.5.59 Instability failure of a sandwich beam hinged on both ends under compressive loading

For the sandwich beam shown in Fig. 2.49, calculate the critical stresses for global and local instability failure assuming thin face sheets and a soft core. Furthermore, sketch the course of the global buckling stress as a function of the beam length L. Given are:

- Geometric dimensions: $L = 2000\,\text{mm}$, $b = 200\,\text{mm}$, $\Delta h^{\text{C}} = 100\,\text{mm}$, $\Delta h^{\text{F}} = 5\,\text{mm}$.
- Material properties of the isotropic core: $E^{\text{C}} = 200\,\text{MPa}$, $\nu^{\text{C}} = 0.4$.
- Material properties of the face sheets: $E^{\text{F}} = 74000\,\text{MPa}$.

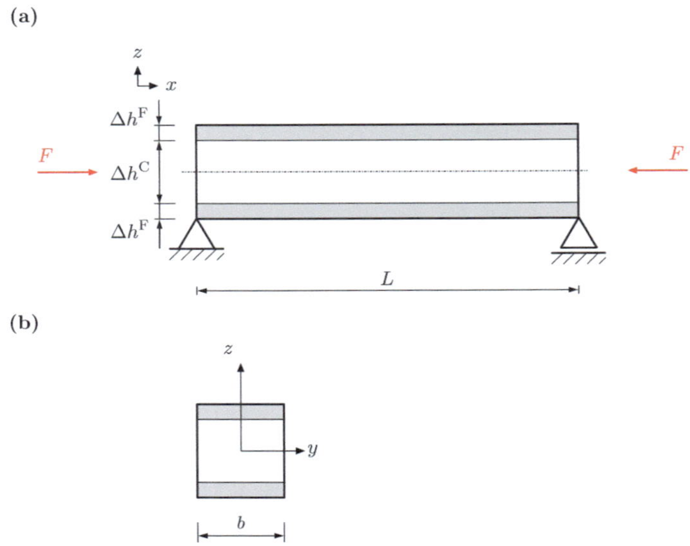

Fig. 2.49 **a** Sandwich beam hinged on both sides under compression; **b** beam cross section

2.6 Sandwich Structures: Optimization

2.6.60 Optimization of a sandwich beam under compressive loading

For the sandwich beam shown in Fig. 2.50, optimize the core and face sheet thicknesses assuming thin face sheets and a soft core.

Given are:

- Geometric dimensions: $L = 2540\,\text{mm}$, $b = 305\,\text{mm}$.
- Material properties of the core: $E^{\text{C}} = 6.8948\,\text{MPa}$, $G^{\text{C}} = 3.4474\,\text{MPa}$, $\varrho^{\text{C}} = 240\,\text{kg/m}^3$.

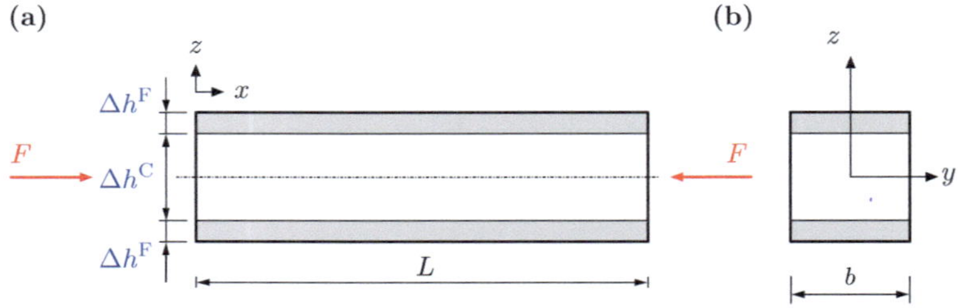

Fig. 2.50 Optimization of a sandwich beam under compressive loading: **a** general configuration and **b** beam cross section

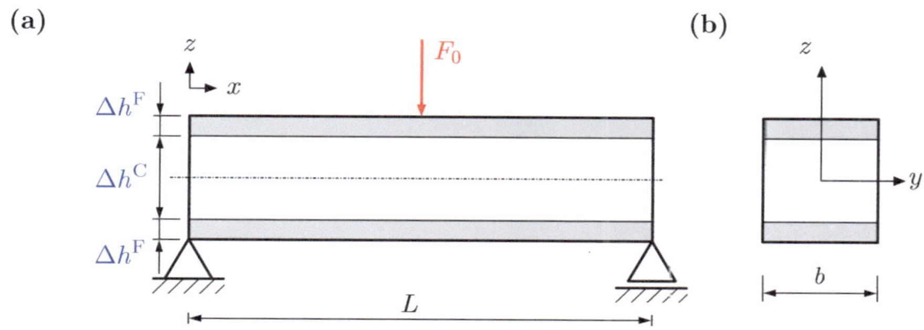

Fig. 2.51 Optimization of a sandwich beam under bending load due to a single force: **a** general configuration and **b** beam cross section

- Material properties of the face sheets: $E^{\mathrm{F}} = 68948\,\mathrm{MPa}$, $\varrho^{\mathrm{F}} = 2691\,\mathrm{kg/m^3}$, $R^{\mathrm{F}}_{\mathrm{p0.2}} = 247\,\mathrm{MPa}$.
- External load: case (a): $F = 2670\,\mathrm{N}$, case (b) $10 \times F = 26700\,\mathrm{N}$.

2.6.61 Optimization of a sandwich beam under bending load due to a single force

For the sandwich beam shown in Fig. 2.51, optimize the core and face sheet thicknesses assuming thin face sheets and a soft core.

Given are:

- Geometric dimensions: $L = 2540\,\mathrm{mm}$, $b = 305\,\mathrm{mm}$.
- Material properties of the core: $E^{\mathrm{C}} = 6.8948\,\mathrm{MPa}$, $G^{\mathrm{C}} = 3.4474\,\mathrm{MPa}$, $\varrho^{\mathrm{C}} = 240\,\mathrm{kg/m^3}$, $\tau^{\mathrm{C}}_{\mathrm{p}} = E^{\mathrm{C}}/50$.
- Material properties of the face sheets: $E^{\mathrm{F}} = 68948\,\mathrm{MPa}$, $\varrho^{\mathrm{F}} = 2691\,\mathrm{kg/m^3}$, $R^{\mathrm{F}}_{\mathrm{p0.2}} = 247\,\mathrm{MPa}$.
- External load: $F_0 = 2667\,\mathrm{N}$.

Furthermore, $r_1 L$ with $r_1 = 0.003$ can be assumed for the maximum deflection.

2.6.62 Optimization of a sandwich beam under bending load due to a distributed load

For the sandwich beam shown in Fig. 2.52, optimize the core and face sheet thicknesses assuming thin face sheets and a soft core.

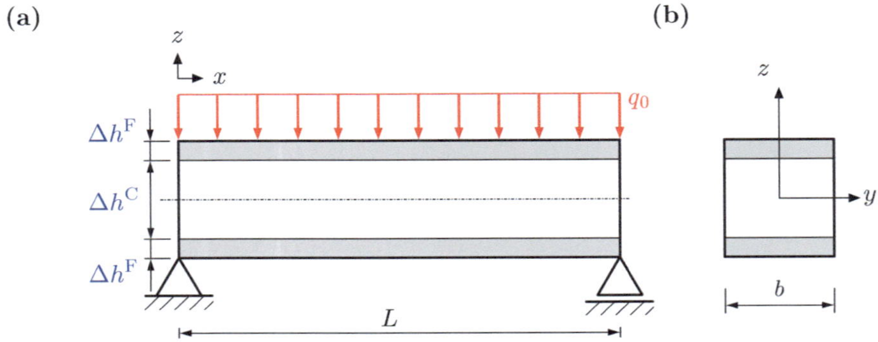

Fig. 2.52 Optimization of a sandwich beam under bending load due to a distributed load: **a** general configuration and **b** beam cross section

Given are:

- Geometric dimensions: $L = 2540\,\text{mm}$, $b = 305\,\text{mm}$.
- Material properties of the core: $E^{\text{C}} = 6.8948\,\text{MPa}$, $G^{\text{C}} = 3.4474\,\text{MPa}$, $\varrho^{\text{C}} = 240\,\text{kg/m}^3$, $\tau_{\text{p}}^{\text{C}} = E^{\text{C}}/50$.
- Material properties of the face sheets: $E^{\text{F}} = 68948\,\text{MPa}$, $\varrho^{\text{F}} = 2691\,\text{kg/m}^3$, $R_{\text{p}0.2}^{\text{F}} = 247\,\text{MPa}$.
- External load: $q_0 = 1.05\,\text{N/mm}$.

Furthermore, $r_1 L$ with $r_1 = 0.003$ can be assumed for the maximum deflection.

2.6.63 Optimization of a homogeneous beam under bending load due to a single force

For the homogeneous beam shown in Fig. 2.53, optimize the dimensions b and h of the cross section for minimum weight. The maximum normal and shear stress, the maximum deflection, and the maximum height-to-width ratio ($h \leq 20b$) in order to avoid instabilities, are to be taken into account as constraints.

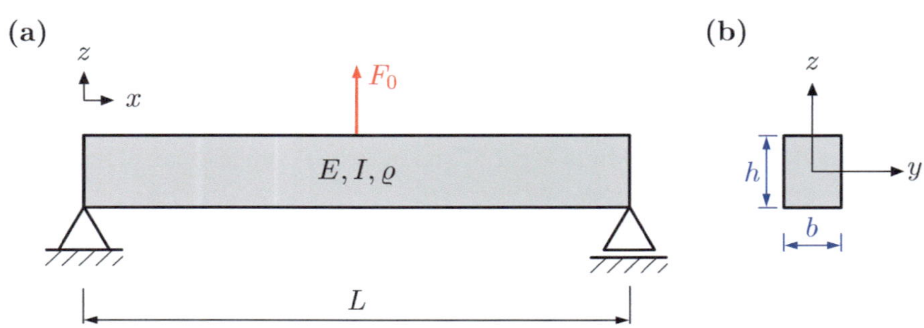

Fig. 2.53 Optimization of a homogeneous beam under bending load due to a single force: **a** general configuration and **b** beam cross section

Given are:

- Geometric dimensions: $L = 2540\,\mathrm{mm}$.
- Material properties of the beam: $E = 68948\,\mathrm{MPa}$, $\varrho = 2691\,\mathrm{kg/m^3}$, $R_{\mathrm{p0.2}} = 247\,\mathrm{MPa}$, $\tau_{\mathrm{p}} = R_{\mathrm{p0.2}}/2$.
- External load: $F_0 = 2667\,\mathrm{N}$.

Furthermore, $r_1 L$ with $r_1 = 0.03$ or $r_1 = 0.003$ can be assumed for the maximum deflection.

References

Javanbakht, Z., Öchsner, A.: Computational Statics Revision Course. Springer, Cham (2018)

Öchsner, A.: Computational Statics and Dynamics: An Introduction Based on the Finite Element Method. Springer, Singapore (2016)

Öchsner, A.: A Project-Based Introduction to Computational Statics. Springer, Cham (2018)

3 Sample Solutions to the Exercises

Abstract

This chapter provides solutions to the exercises. Not only are final solutions provided, but also detailed intermediate steps so that the solutions can be understood.

3.1 Basics of Mechanics of Materials

2.1.1 **Simplified model of a tower under dead weight**

The load results from the vertical force equilibrium as (see Fig. 3.1)

$$N_x(x) = -F_{\text{G}} = -mg = -\varrho V g = -\varrho g A(L - x), \tag{3.1}$$

or as a distributed load:

$$p_x(x) = -\frac{\mathrm{d}N_x(x)}{\mathrm{d}x} = -\varrho g A = p_0 = \text{constant}. \tag{3.2}$$

Alternatively, Eq. (3.2) can also be derived by normalizing the total dead weight of the tower using the length L[1]:

$$p_x(x) = -\frac{F_{\text{G}}}{L} = -\frac{mg}{L} = -\varrho g A = p_0 = \text{constant}. \tag{3.3}$$

[1] It should be noted that the distributed load $p_x(x)$ acts against the positive x-direction.

A. Öchsner, *Exercises and Solutions in Lightweight Design*,
https://doi.org/10.1007/978-3-658-49333-2_3

Fig. 3.1 Internal reaction (N_x) in the simplified model of the tower under its dead weight

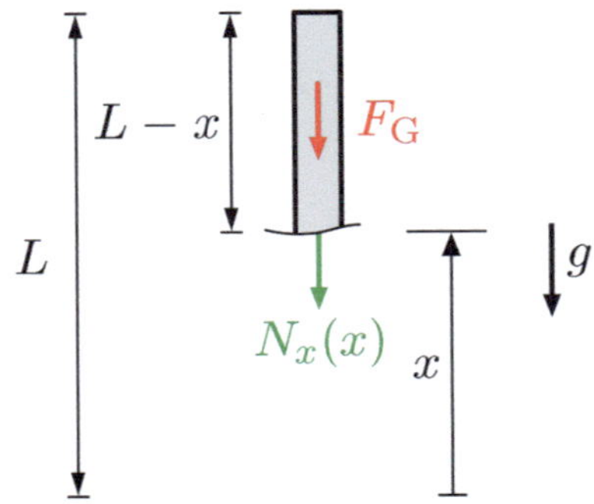

Due to the constant tensile stiffness EA and the constant distributed load p_0, the general solutions according to Eqs. $(2.2)_{\text{TB}}$ and $(2.3)_{\text{TB}}$ can be used:

$$N_x(x) = EA\frac{\mathrm{d}u_x(x)}{\mathrm{d}x} = -p_0 x + c_1. \tag{3.4}$$

$$u_x(x) = \frac{1}{EA}\left(-\frac{1}{2}p_0 x^2 + c_1 x + c_2\right). \tag{3.5}$$

Using the boundary conditions $u_x(x=0)=0$ and $N_x(x=L)=0$, the integration constants $c_2=0$ and $c_1=p_0L=-\varrho gAL$ are obtained from Eq. (3.5). Thus, the normal force and displacement distributions are:

$$N_x(x) = -\varrho gA(L-x), \tag{3.6}$$

$$u_x(x)\frac{1}{EA}\left(+\tfrac{1}{2}\varrho gAx^2 - \varrho gALx\right). \tag{3.7}$$

The stress distribution results from the normal force distribution as:

$$\sigma_x(x) = \frac{N_x(x)}{A} = -\varrho g(L-x). \tag{3.8}$$

The length reduced by the dead weight is:

$$L' = L + u_x(L) = L\left(1 - \frac{\varrho gL}{2E}\right). \tag{3.9}$$

The maximum length L_{max}, if a given stress σ_{max} in the foundation must not be exceeded, results from the condition

$$|\sigma_{\text{max}}| = |\sigma_x(x=0)| = \varrho gL_{\text{max}}, \tag{3.10}$$

or

$$L_{\text{max}} = \frac{\sigma_{\text{max}}}{\varrho g}. \tag{3.11}$$

2.1.2 Course of the bending line and maximum stress for a cantilever beam with different loads

Due to the constant bending stiffness (EI_y) and the constant distributed load (part (a): $q_0 = 0$, part (b): $-q_0$, since in the negative z-direction), the shear force, bending moment and rotation distributions can generally be calculated according to Eqs. $(2.12)_{\text{TB}}$–$(2.15)_{\text{TB}}$:

$$Q_z(x) = -(-q_0)x - c_1, \tag{3.12}$$

$$M_y(x) = -\frac{(-q_0)x^2}{2} - c_1 x - c_2, \tag{3.13}$$

$$\varphi_y(x) = -\frac{1}{EI_y}\left(\frac{(-q_0)x^3}{6} + \frac{c_1 x^2}{2} + c_2 x + c_3\right), \tag{3.14}$$

$$u_z(x) = \frac{1}{EI_y}\left(\frac{(-q_0)x^4}{24} + \frac{c_1 x^3}{6} + \frac{c_2 x^2}{2} + c_3 x + c_4\right), \tag{3.15}$$

where in subtask (a) $q_0 = 0$ must be set.

Load case (a): Using the boundary conditions

$$u_z(x = 0) = 0, \tag{3.16}$$

$$\varphi_y(x = 0) = 0, \tag{3.17}$$

$$M_y(x = L) = 0, \tag{3.18}$$

$$Q_z(x = L) = -F_0, \tag{3.19}$$

the integration constants are obtained as:

$$(3.16) \text{ in } (3.15): \quad c_4 = 0, \tag{3.20}$$

$$(3.17) \text{ in } (3.14): \quad c_3 = 0, \tag{3.21}$$

$$(3.19) \text{ in } (3.12): \quad c_1 = F_0, \tag{3.22}$$

$$(3.18) \text{ in } (3.13): \quad c_2 = -F_0 L. \tag{3.23}$$

With these integration constants, the distributions of the deflection and the bending moment are as follows (see also Fig. 3.2a):

$$u_z(x) = \frac{F_0 L^3}{EI}\left(\frac{1}{6}\left[\frac{x}{L}\right]^3 - \frac{1}{2}\left[\frac{x}{L}\right]^2\right), \tag{3.24}$$

$$M_y(x) = F_0 L\left(-\left[\frac{x}{L}\right] + 1\right). \tag{3.25}$$

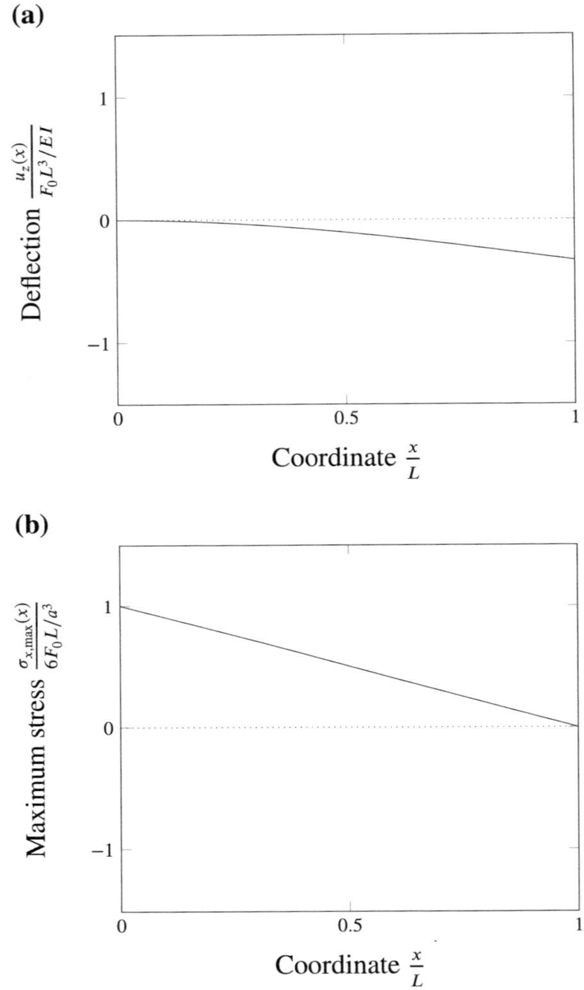

Fig. 3.2 Course of the **a** bending line and **b** maximum stress for a cantilever beam with a point load

The maximum bending stress on the top side of the beam (i.e., at $z = +a/2$) results from the bending moment distribution (see also Fig. 3.2b):

$$\sigma_{x,\max}(x) = \frac{M_y(x)}{I_y} \times z_{\max} = \frac{M_y(x)}{\frac{a^4}{12}} \times \frac{a}{2} = \frac{6F_0L}{a^3}\left(-\left[\frac{x}{L}\right] + 1\right). \tag{3.26}$$

Load case (b): Using the boundary conditions

$$u_z(x=0)=0, \tag{3.27}$$
$$\varphi_y(x=0)=0, \tag{3.28}$$
$$M_y(x=L)=0, \tag{3.29}$$
$$Q_z(x=L)=0, \tag{3.30}$$

the integration constants are obtained as:

$$(3.27) \text{ in } (3.15): \quad c_4=0, \tag{3.31}$$
$$(3.28) \text{ in } (3.14): \quad c_3=0, \tag{3.32}$$
$$(3.30) \text{ in } (3.12): \quad c_1=q_0L, \tag{3.33}$$
$$(3.29) \text{ in } (3.13): \quad c_2=-\tfrac{q_0L^2}{2}. \tag{3.34}$$

With these integration constants, the distributions of the deflection and the bending moment are as follows (see also Fig. 3.3a):

$$u_z(x)=-\frac{q_0L^4}{24EI}\left(\left[\frac{x}{L}\right]^4-4\left[\frac{x}{L}\right]^3+6\left[\frac{x}{L}\right]^2\right), \tag{3.35}$$

$$M_y(x)=\frac{q_0L^2}{2}\left(\left[\frac{x}{L}\right]^2-2\left[\frac{x}{L}\right]+1\right). \tag{3.36}$$

The maximum bending stress on the top side of the beam (i.e., at $z=+a/2$) results from the bending moment distribution (see also Fig. 3.3b):

$$\sigma_{x,\max}(x)=\frac{M_y(x)}{I_y}\times z_{\max}=\frac{M_y(x)}{\frac{a^4}{12}}\times\frac{a}{2}=\frac{3q_0L^2}{a^3}\left(\left[\frac{x}{L}\right]^2-2\left[\frac{x}{L}\right]+1\right). \tag{3.37}$$

2.1.3 **Course of bending line, stress and strain for a bi-material cantilever beam**

The use of two different materials requires that the beam be divided into two sections, see Fig. 3.4.

For each of these sections, the general solutions according to Eqs. $(2.12)_{\text{TB}}$–$(2.15)_{\text{TB}}$ are applied separately:

$$Q_{z_\text{I}}(x_\text{I})=-c_1, \tag{3.38}$$
$$M_{y_\text{I}}(x_\text{I})=-c_1x_\text{I}-c_2, \tag{3.39}$$
$$\varphi_{y_\text{I}}(x_\text{I})=-\frac{1}{EI_y}\left(\frac{c_1x_\text{I}^2}{2}+c_2x_\text{I}+c_3\right), \tag{3.40}$$
$$u_{z_\text{I}}(x_\text{I})=\frac{1}{EI_y}\left(\frac{c_1x_\text{I}^3}{6}+\frac{c_2x_\text{I}^2}{2}+c_3x_\text{I}+c_4\right), \tag{3.41}$$

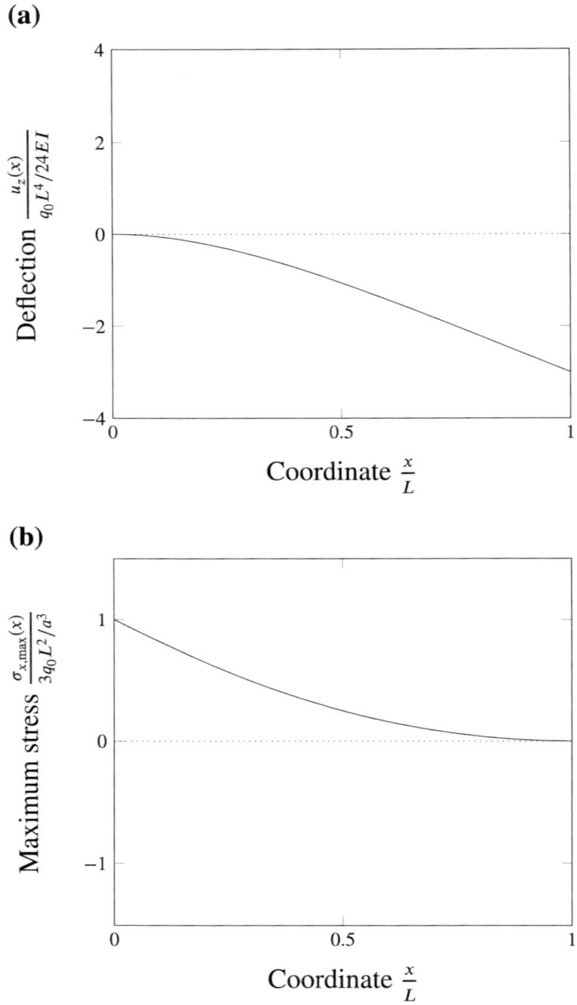

Fig. 3.3 Course of the **a** bending line and **b** maximum stress for a cantilever beam with a constant distributed load

or for the second section

$$Q_{z\mathrm{II}}(x_{\mathrm{II}}) = -c_5, \tag{3.42}$$

$$M_{y\mathrm{II}}(x_{\mathrm{II}}) = -c_5 x_{\mathrm{II}} - c_6, \tag{3.43}$$

$$\varphi_{y\mathrm{II}}(x_{\mathrm{II}}) = -\frac{1}{EI_y}\left(\frac{c_5 x_{\mathrm{II}}^2}{2} + c_6 x_{\mathrm{II}} + c_7\right), \tag{3.44}$$

$$u_{z\mathrm{II}}(x_{\mathrm{II}}) = \frac{1}{EI_y}\left(\frac{c_5 x_{\mathrm{II}}^3}{6} + \frac{c_6 x_{\mathrm{II}}^2}{2} + c_7 x_{\mathrm{II}} + c_8\right), \tag{3.45}$$

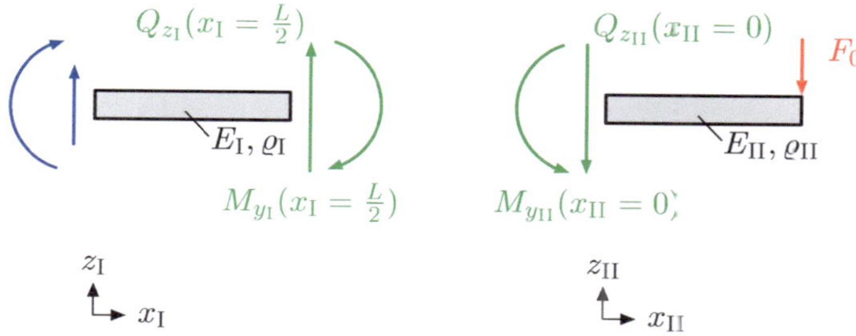

Fig. 3.4 Division of the bi-material cantilever beam into two sections

where $q_0 = 0$ has already been used. Using the boundary conditions

$$u_{z_I}(x_I = 0) = 0, \quad M_{y_{II}}\left(x_{II} = \tfrac{L}{2}\right) = 0, \tag{3.46}$$

$$\varphi_{y_I}(x_I = 0) = 0, \quad Q_{z_{II}}\left(x_{II} = \tfrac{L}{2}\right) = -F_0, \tag{3.47}$$

and accordingly the transition conditions at the location $x = \frac{L}{2}$ (or $x_I = \frac{L}{2}$ respectively $x_{II} = 0$)

$$u_{z_I}\left(x_I = \tfrac{L}{2}\right) = u_{z_{II}}(x_{II} = 0), \tag{3.48}$$

$$\varphi_{y_I}\left(x_I = \tfrac{L}{2}\right) = \varphi_{y_{II}}(x_{II} = 0), \tag{3.49}$$

$$Q_{z_I}\left(x_I = \tfrac{L}{2}\right) = Q_{z_{II}}(x_{II} = 0), \tag{3.50}$$

$$M_{y_I}\left(x_I = \tfrac{L}{2}\right) = M_{y_{II}}(x_{II} = 0), \tag{3.51}$$

the eight integration constants are obtained as:

$$c_1 = F_0, c_2 = -F_0 L, c_3 = c_4 = 0, c_5 = F_0, c_6 = -\tfrac{F_0 L}{2}, \tag{3.52}$$

$$c_7 = -\tfrac{3L^2 E_{II} F_0}{8E_I}, c_8 = -\tfrac{5L^3 E_{II} F_0}{48 E_I}. \tag{3.53}$$

With these integration constants, the section-wise bending lines are (see Fig. 3.5a):

$$u_{z_I}(x_I) = \frac{F_0 L^3}{E_I I_y}\left(\frac{1}{6}\left[\frac{x_I}{L}\right]^3 - \frac{1}{2}\left[\frac{x_I}{L}\right]^2\right), \tag{3.54}$$

$$u_{z_{II}}(x_{II}) = \frac{F_0 L^3}{E_{II} I_y}\left(\frac{1}{6}\left[\frac{x_{II}}{L}\right]^3 - \frac{1}{4}\left[\frac{x_{II}}{L}\right]^2 - \frac{3}{8}\frac{E_{II}}{E_I}\left[\frac{x_{II}}{L}\right]^1 - \frac{5}{48}\frac{E_{II}}{E_I}\right). \tag{3.55}$$

From the Eqs. (3.39) and (3.43) the distributions of the bending moments are given as:

$$M_{y_I}(x_I) = F_0\,(L - x_I)\,, \tag{3.56}$$

$$M_{y_{II}}(x_{II}) = F_0\left(\frac{L}{2} - x_{II}\right), \tag{3.57}$$

respectively the stresses on the top of the beam, i.e. $z = +a/2$ (see Fig. 3.5b):

$$\sigma_{z\mathrm{I},\max}(x_\mathrm{I}) = \frac{M_{y_\mathrm{I}}(x_\mathrm{I})}{I_{y_\mathrm{I}}} \times z_{\mathrm{I},\max} = \frac{F_0 L a}{2 I_y}\left(1 - \left[\frac{x_\mathrm{I}}{L}\right]\right), \tag{3.58}$$

$$\sigma_{z\mathrm{II},\max}(x_\mathrm{II}) = \frac{M_{y_\mathrm{II}}(x_\mathrm{II})}{I_{y_\mathrm{II}}} \times z_{\mathrm{II},\max} = \frac{F_0 L a}{2 I_y}\left(\frac{1}{2} - \left[\frac{x_\mathrm{II}}{L}\right]\right). \tag{3.59}$$

Application of Hooke's law ($\sigma = E\varepsilon$) yields the strains (see Fig. 3.5c):

$$\varepsilon_{z\mathrm{I},\max}(x_\mathrm{I}) = \frac{F_0 L a}{2 E_\mathrm{I} I_y}\left(1 - \left[\frac{x_\mathrm{I}}{L}\right]\right), \tag{3.60}$$

$$\varepsilon_{z\mathrm{II},\max}(x_\mathrm{II}) = \frac{F_0 L a}{2 E_\mathrm{II} I_y}\left(\frac{1}{2} - \left[\frac{x_\mathrm{II}}{L}\right]\right). \tag{3.61}$$

2.1.4 **Comparison of the maximum deflections of a cantilever beam with constant line load according to different beam theories**

The general configuration of the cantilever beam with constant distributed load is shown in Fig. 3.6.

The reference solution, i.e. the maximum deflection according to Euler-Bernoulli, can be taken from the solution to the Exercise 2.1.2 for $x = L$:

$$u_z(x = L) = -\frac{q_0 L^4}{8 E I_y}. \tag{3.62}$$

To determine the maximum beam deflection according to Timoshenko's theory, the general solutions according to Eqs. $(2.22)_\mathrm{TB}$–$(2.25)_\mathrm{TB}$ can be used (the distributed load and the material and geometric properties are constant):

$$Q_z(x) = -\left((-q_0)x + c_1\right), \tag{3.63}$$

$$M_y(x) = -\left(\frac{(-q_0)x^2}{2} + c_1 x + c_2\right) - \frac{q_z E I_y}{k_\mathrm{s} A G}, \tag{3.64}$$

$$\phi_y(x) = -\frac{1}{E I_y}\left(\frac{(-q_0)x^3}{6} + c_1\frac{x^2}{2} + c_2 x + c_3\right) - \frac{(-q_0)x}{k_\mathrm{s} A G} - \frac{c_1}{k_\mathrm{s} A G}, \tag{3.65}$$

$$u_z(x) = \frac{1}{E I_y}\left(\frac{(-q_0)x^4}{24} + c_1\frac{x^3}{6} + c_2\frac{x^2}{2} + c_3 x + c_4\right), \tag{3.66}$$

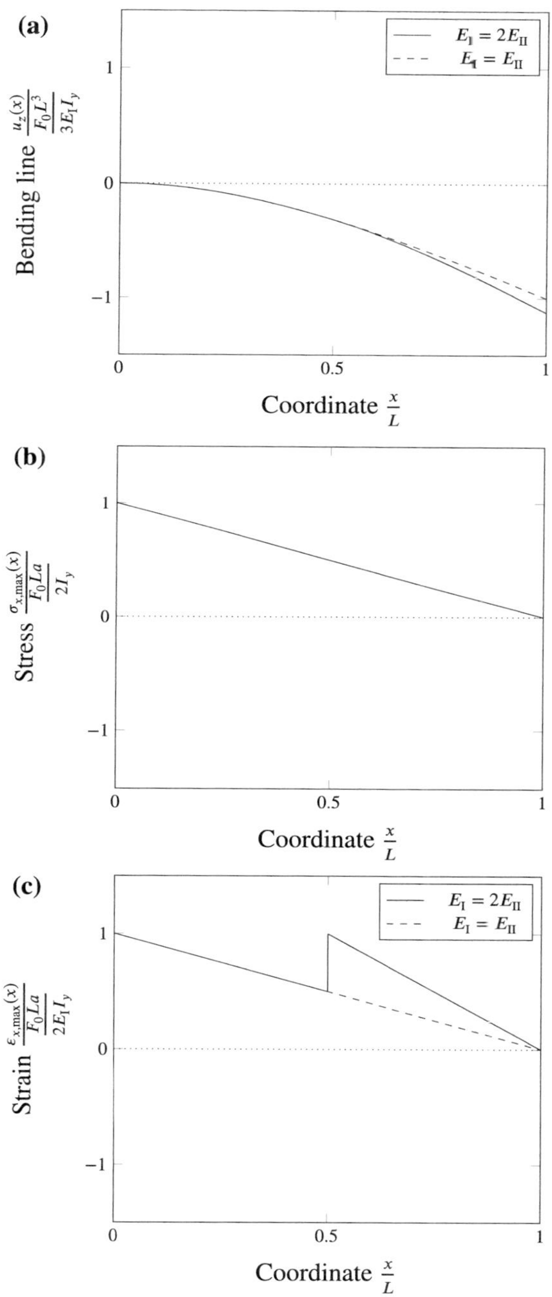

Fig. 3.5 Course of **a** bending line, **b** stress (no influence of E_i) and **c** strain for the bi-material cantilever beam

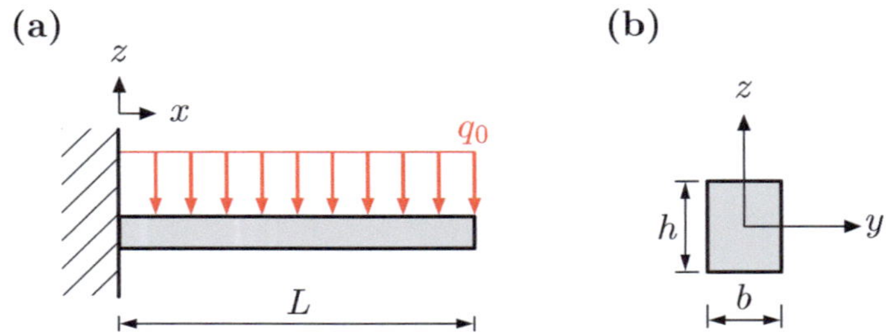

Fig. 3.6 General configuration of the cantilever beam with constant distributed load: **a** Geometric boundary conditions and external load; **b** beam cross section

where the substitution $q_z \rightarrow -q_0$ (loading opposite to the positive z-direction) was also performed. Using the boundary conditions

$$u_z(x=0)=0, \tag{3.67}$$

$$\varphi_y(x=0)=0, \tag{3.68}$$

$$M_y(x=L)=0, \tag{3.69}$$

$$Q_z(x=L)=0, \tag{3.70}$$

the integration constants are obtained as summarized in Table 3.1.
Using these integration constants, the deflection for the Timoshenko beam at the location $x = L$ is:

$$u_z(L)|_{\mathrm{T}} = \frac{1}{EI_y}\left(\frac{(-q_0)x^4}{24} + \underbrace{(q_0L)}_{c_1}\frac{L^3}{6} + \underbrace{(-\frac{q_0L^2}{2} + \frac{q_0EI_y}{k_sAG})}_{c_2}\frac{L^2}{2} + \underbrace{(-\frac{q_0LEI_y}{k_sAG})}_{c_3}L\right), \tag{3.71}$$

Table 3.1 Integration constants and normalized deflection according to the various beam theories for a cantilever with a constant line load (q_0 in negative z-direction)

Theory	c_1	c_2	c_3	c_4	$\frac{u_z(L)}{-\frac{q_0L^4}{8EI_y}}$
Euler-Bernoulli	q_0L	$-\frac{q_0L^2}{2}$	0	0	1
Timoshenko	q_0L	$-\frac{q_0L^2}{2} + \frac{q_0EI_y}{k_sAG}$	$-\frac{q_0LEI_y}{k_sAG}$	0	$1 + \frac{4(1+\nu)}{5}\left(\frac{h}{L}\right)^2$
Levinson	q_0L	$-\frac{q_0L^2}{2} + \frac{6q_0EI_y}{5AG}$	$-\frac{3q_0LEI_y}{2AG}$	0	$1 + \frac{6(1+\nu)}{5}\left(\frac{h}{L}\right)^2$

respectively by means of the following relationships

$$k_s = \frac{5}{6}, \tag{3.72}$$

$$G = \frac{E}{2(1+\nu)}, \tag{3.73}$$

$$I_y = \frac{1}{12}bh^3 = \frac{1}{12} \times \underbrace{bh}_{A} \times h^2 = \frac{1}{12}Ah^2, \tag{3.74}$$

after a short transformation in normalized representation for the Timoshenko beam:

$$\frac{u_{z,\max}}{-\frac{q_0 L^4}{EI_y}} = \frac{1}{8} + \frac{1+\nu}{10}\left(\frac{h}{L}\right)^2. \tag{3.75}$$

To determine the maximum beam deflection according to Levinson's theory, the general solutions according to Eqs. $(2.31)_{\text{TB}}$–$(2.34)_{\text{TB}}$ can be used (the distributed load and the material and geometric properties are constant):

$$Q_z(x) = -\left((-q_0)x + c_1\right), \tag{3.76}$$

$$M_y(x) = -\left(\frac{(-q_0)x^2}{2} + c_1 x + c_2\right) - \frac{6(-q_0)EI_y}{5AG}, \tag{3.77}$$

$$\phi_y(x) = -\frac{1}{EI_y}\left(\frac{(-q_0)x^3}{6} + c_1\frac{x^2}{2} + c_2 x + c_3\right) - \frac{(-q_0)x}{\frac{2}{3}AG} - \frac{c_1}{\frac{2}{3}AG}. \tag{3.78}$$

$$u_z(x) = \frac{1}{EI_y}\left(\frac{(-q_0)x^4}{24} + c_1\frac{x^3}{6} + c_2\frac{x^2}{2} + c_3 x + c_4\right), \tag{3.79}$$

where the substitution $q_z \rightarrow -q_0$ (loading opposite to the positive z-direction) was also performed. Using the boundary conditions

$$u_z(x=0) = 0, \tag{3.80}$$

$$\varphi_y(x=0) = 0, \tag{3.81}$$

$$M_y(x=L) = 0, \tag{3.82}$$

$$Q_z(x=L) = 0, \tag{3.83}$$

the integration constants are obtained as summarized in Table 3.1. Using the same procedure as for the Timoshenko beam, the normalized deflection for the Levinson beam is finally obtained as:

$$\frac{u_{z,\max}}{-\frac{q_0 L^4}{EI_y}} = \frac{1}{8} + \frac{3(1+\nu)}{20}\left(\frac{h}{L}\right)^2. \tag{3.84}$$

2.1.5 Axial second moment of area for a circular ring

For a full circular cross section with diameter D or radius R, the axial second moment of area is generally:

$$I = \frac{\pi}{64} D^4 = \frac{\pi}{4} R^4. \tag{3.85}$$

- Thick wall ring I:

$$I = \frac{\pi}{4}\left(r_\mathrm{a}^4 - r_\mathrm{i}^4\right) = \frac{\pi}{4}\left(r_\mathrm{a}^2 + r_\mathrm{i}^2\right)\left(r_\mathrm{a}^2 - r_\mathrm{i}^2\right) \tag{3.86}$$

$$= \frac{\pi}{4}\left(r_\mathrm{a}^2 + r_\mathrm{i}^2\right)\left(r_\mathrm{a} - r_\mathrm{i}\right)\left(r_\mathrm{a} + r_\mathrm{i}\right). \tag{3.87}$$

With $r_\mathrm{m} = \frac{1}{2}\left(r_\mathrm{a} + r_\mathrm{i}\right)$, $s = \left(r_\mathrm{a} - r_\mathrm{i}\right)$ and $r_\mathrm{a} = r_\mathrm{m} + \frac{s}{2}$ or $r_\mathrm{i} = r_\mathrm{m} - \frac{s}{2}$ respectively

$$\left(r_\mathrm{a}^2 + r_\mathrm{i}^2\right) = \left(r_\mathrm{m} + \frac{s}{2}\right)^2 + \left(r_\mathrm{m} - \frac{s}{2}\right)^2 \tag{3.88}$$

$$= r_\mathrm{m}^2\left(1 + \frac{s}{2r_\mathrm{m}}\right)^2 + r_\mathrm{m}^2\left(1 - \frac{s}{2r_\mathrm{m}}\right)^2 = 2r_\mathrm{m}^2\left(1 + \frac{s^2}{4r_\mathrm{m}^2}\right), \tag{3.89}$$

finally follows for the axial second moment of area

$$I = \pi r_\mathrm{m}^3\, s\left(1 + \frac{s^2}{4r_\mathrm{m}^2}\right) = \frac{\pi}{8} d_\mathrm{m}^3\, s\left(1 + \frac{s^2}{d_\mathrm{m}^2}\right), \tag{3.90}$$

where $d_\mathrm{m} = 2r_\mathrm{m}$ denotes the mean diameter.

- Thin wall ring I:

For $s \ll r_\mathrm{m}$ in Eq. (3.90) the expression $\frac{s^2}{4r_\mathrm{m}^2} \ll 1$ can be neglected compared to "1" and the following results approximately:

$$I \approx \pi r_\mathrm{m}^3\, s = \frac{\pi}{8} d_\mathrm{m}^3\, s. \tag{3.91}$$

In general, using the true value a and the approximate value $\overline{a}$, the relative error can be set as follows ($a \neq 0$):

$$\left|\frac{\Delta a}{a}\right| = \frac{|a - \overline{a}|}{|a|}. \tag{3.92}$$

For a relative error of less than 1% for the simplification, our example yields:

$$\frac{\pi r_{\mathrm{m}}^3 s \times \dfrac{s^2}{4r_{\mathrm{m}}^2}}{\pi r_{\mathrm{m}}^3 s \left(1 + \dfrac{s^2}{4r_{\mathrm{m}}^2}\right)} \leq 0.01 \quad \Rightarrow \frac{r_{\mathrm{m}}^2}{s^2} \geq \frac{1 - 0.01}{4 \times 0.01} = 24.75 \tag{3.93}$$

$$\Rightarrow \quad \frac{r_{\mathrm{m}}}{s} \geq 4.97. \tag{3.94}$$

- Cross-sectional area A:

For a full circle cross section with radius R the cross-sectional area is generally:

$$A = \pi\ R^2. \tag{3.95}$$

Thus, for the tube with inner radius r_{i} and outer radius r_{o}, the following relationship can be applied:

$$A = \pi\ \left(r_{\mathrm{o}}^2 - r_{\mathrm{i}}^2\right) = \pi\ (r_{\mathrm{o}} - r_{\mathrm{i}})\ (r_{\mathrm{o}} + r_{\mathrm{i}}) \tag{3.96}$$

$$= 2\pi r_{\mathrm{m}} s = \pi d_{\mathrm{m}} s \quad \text{(exact)}, \tag{3.97}$$

or

$$A = \pi\ (r_{\mathrm{o}} + r_{\mathrm{i}})\ s, \tag{3.98}$$

respectively for $s \ll r_{\mathrm{m}}$ with $r_{\mathrm{o}} \approx r_{\mathrm{i}}$:

$$A \approx 2\pi r_{\mathrm{i}} s \approx 2\pi r_{\mathrm{o}} s. \tag{3.99}$$

2.1.6 **Axial second moment of area for a square box profile**

For a square solid cross section with side length a, the axial second moment of area is generally:

$$I = \frac{1}{12} a^4. \tag{3.100}$$

- Thick profile I:

$$I = \frac{1}{12}\left(a_{\mathrm{a}}^4 - a_{\mathrm{i}}^4\right) = \frac{1}{12}\left(a_{\mathrm{a}}^2 + a_{\mathrm{i}}^2\right)\left(a_{\mathrm{a}}^2 - a_{\mathrm{i}}^2\right) \tag{3.101}$$

$$= \frac{1}{12}\left(a_{\mathrm{a}}^2 + a_{\mathrm{i}}^2\right)(a_{\mathrm{a}} - a_{\mathrm{i}})\,(a_{\mathrm{a}} + a_{\mathrm{i}})\,. \tag{3.102}$$

With $a_\mathrm{m} = \frac{1}{2}\left(a_\mathrm{a} + a_\mathrm{i}\right)$, $s = \frac{1}{2}\left(a_\mathrm{a} - a_\mathrm{i}\right)$ and $a_\mathrm{a} = a_\mathrm{m} + s$ or $a_\mathrm{i} = a_\mathrm{m} - s$ respectively

$$\left(a_\mathrm{a}^2 + a_\mathrm{i}^2\right) = (a_\mathrm{m} + s)^2 + (a_\mathrm{m} - s)^2 \tag{3.103}$$

$$= a_\mathrm{m}^2\left(1 + \frac{s}{a_\mathrm{m}}\right)^2 + a_\mathrm{m}^2\left(1 - \frac{s}{a_\mathrm{m}}\right)^2 = 2a_\mathrm{m}^2\left(1 + \frac{s^2}{a_\mathrm{m}^2}\right), \tag{3.104}$$

finally follows for the axial second moment of area

$$I = \frac{2}{3}a_\mathrm{m}^3\, s\left(1 + \frac{s^2}{a_\mathrm{m}^2}\right). \tag{3.105}$$

- Thin profile I:

For $s \ll a_\mathrm{m}$ in Eq. (3.105) the expression $\frac{s^2}{a_\mathrm{m}^2} \ll 1$ can be neglected compared to "1" and the following results approximately:

$$I \approx \frac{2}{3}a_\mathrm{m}^3\, s. \tag{3.106}$$

For a relative error of less than 1% for the simplification, our example yields:

$$\frac{\frac{2}{3}\,a_\mathrm{m}^3\, s \times \dfrac{s^2}{a_\mathrm{m}^2}}{\frac{2}{3}\,a_\mathrm{m}^3\, s\left(1 + \dfrac{s^2}{a_\mathrm{m}^2}\right)} \leq 0.01 \quad \Rightarrow \frac{a_\mathrm{m}^2}{s^2} \geq \frac{1 - 0.01}{0.01} = 99 \tag{3.107}$$

$$\Rightarrow \quad \frac{a_\mathrm{m}}{s} \geq 9.95. \tag{3.108}$$

- Cross-sectional area A:

For a square solid cross section with side length a the cross-sectional area is generally:

$$A = a^2. \tag{3.109}$$

Thus, for the square box profile with outer dimension a_o and inner dimension a_i, the following relationship can be applied:

$$A = a_\mathrm{o}^2 - a_\mathrm{i}^2 = (a_\mathrm{o} - a_\mathrm{i})\,(a_\mathrm{o} + a_\mathrm{i}) \tag{3.110}$$

$$= 4a_\mathrm{m}s \quad \text{(exact)}. \tag{3.111}$$

2.1.7 **General solution of the differential equations of the Timoshenko beam using a computer algebra systems**

The solution for the deflection using Maxima can be given according to the order of the polynomial in x as follows:

$$\begin{aligned} u(x) = {} & \frac{q}{24EI} \times x^4 \\ & + \underbrace{\left(-\frac{kAG \, \frac{\mathrm{d}}{\mathrm{d}x} \, \mathrm{u}(x)\Big|_{x=0} + \phi(0)kAG}{24EI} - \frac{kAG \, \frac{\mathrm{d}}{\mathrm{d}x} \, \mathrm{u}(x)\Big|_{x=0}}{8EI} - \frac{\phi(0)kAG}{8EI} \right)}_{C1/6} \times x^3 \\ & + \underbrace{\left(-\frac{2kAG \, \frac{\mathrm{d}}{\mathrm{d}x} \phi(x)\Big|_{x=0} + 2q}{24kAG} - \frac{kAG \, \frac{\mathrm{d}}{\mathrm{d}x} \phi(x)\Big|_{x=0} + q}{6kAG} - \frac{\frac{\mathrm{d}}{\mathrm{d}x} \phi(x)\Big|_{x=0}}{4} - \frac{q}{4kAG} \right)}_{C2/2} \times x^2 \\ & + \underbrace{\left(\frac{\mathrm{d}}{\mathrm{d}x} \, \mathrm{u}(x)\Big|_{x=0} \right)}_{C3} \times x^1 \\ & + \underbrace{\mathrm{u}(0)}_{C4} . \end{aligned} \tag{3.112}$$

Comparing the coefficients with respect to the solution according to Maple® and simplifying the bracket expressions for $C1/6$ and $C2/2$ finally yields the following identities:

$$C1 = -\frac{kAG \, \frac{\mathrm{d}u}{\mathrm{d}x}\big|_{x=0}}{EI} - \frac{kAG\phi(0)}{EI}, \tag{3.113}$$

$$C2 = -\left.\frac{\mathrm{d}\phi}{\mathrm{d}x}\right|_{x=0} - \frac{q}{kAG}, \tag{3.114}$$

$$C3 = \left.\frac{\mathrm{d}u(x)}{\mathrm{d}x}\right|_{x=0}, \tag{3.115}$$

$$C4 = u(0). \tag{3.116}$$

2.1.8 **Maximum difference between von Mises and Tresca**

From Figs. 2.21_{TB} and 2.23_{TB} the maximum absolute error between von Mises and Tresca in the σ-τ-stress plane for $\sigma = 0$ is:

$$\left| \frac{1}{\sqrt{3}} k_{\mathrm{t}} - \frac{1}{2} k_{\mathrm{t}} \right| = \frac{2 - \sqrt{3}}{2\sqrt{3}} k_{\mathrm{t}} = 0.07735 k_{\mathrm{t}} \approx 7.7\% k_{\mathrm{t}}. \tag{3.117}$$

Maximum relative error (option 1, i.e., von Mises assumed to be the true value):

$$\left|\frac{\frac{1}{\sqrt{3}}k_t - \frac{1}{2}k_t}{\frac{1}{\sqrt{3}}k_t}\right| = \frac{2-\sqrt{3}}{2} = 0.13397 \approx 13.4\%. \tag{3.118}$$

Maximum relative error (option 2, i.e., Tresca assumed to be the true value):

$$\left|\frac{\frac{1}{2}k_t - \frac{1}{\sqrt{3}}k_t}{\frac{1}{2}k_t}\right| = \left|\frac{\sqrt{3}-2}{\sqrt{3}}\right| = 0.15470 \approx 15.5\%. \tag{3.119}$$

2.1.9 **Bar under tensile and torsional loading**

The stress components for the basic load cases of tension and torsion are obtained according to Eqs. $(2.4)_{\text{TB}}$ and $(2.8)_{\text{TB}}$ at the surface of the bar as:

$$\sigma = \frac{N_x(x)}{A} = \frac{F_0}{\frac{\pi}{4}h^2} = \frac{4F_0}{\pi h^2}, \tag{3.120}$$

$$\tau = \frac{M_x(x)}{I_p} \times r_{\max} = \frac{M_t}{\frac{\pi h^4}{32}} \times \frac{h}{2} = \frac{16M_t}{\pi h^3}. \tag{3.121}$$

The equivalent stress according to von Mises for a stress state with the components σ and τ is given by Eq. $(2.37)_{\text{TB}}$ as:

$$\sigma_{\text{eff}} = \sqrt{\sigma^2 + 3\tau^2} = \frac{4}{\pi} \times \sqrt{\left(\frac{F_0}{h^2}\right)^2 + 3\left(\frac{4M_t}{h^3}\right)^2}. \tag{3.122}$$

3.2 Lightweight Design Concepts: Material Selection

2.2.10 **Finite element calculation of a stepped tensile bar**

Information on the individual steps of a finite element analysis can be found in Öchsner (2016, 2018); Javanbakht und Öchsner (2017). The discretization of the stepped tensile bar is shown in Fig. 3.7. Node and element numbers are also given there.

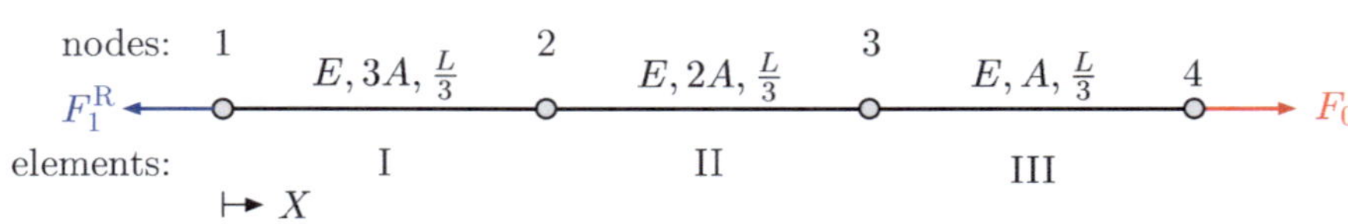

Fig. 3.7 Discretization of the stepped tensile bar

The stiffness matrices of the individual elements are:

$$\boldsymbol{K}_{\mathrm{I}}^{\mathrm{e}} = \frac{E3A}{\frac{1}{3}L}\begin{bmatrix} 1 & -1 \\ -1 & 1 \end{bmatrix}, \tag{3.123}$$

$$\boldsymbol{K}_{\mathrm{II}}^{\mathrm{e}} = \frac{E2A}{\frac{1}{3}L}\begin{bmatrix} 1 & -1 \\ -1 & 1 \end{bmatrix}, \tag{3.124}$$

$$\boldsymbol{K}_{\mathrm{III}}^{\mathrm{e}} = \frac{EA}{\frac{1}{3}L}\begin{bmatrix} 1 & -1 \\ -1 & 1 \end{bmatrix}. \tag{3.125}$$

This results in the global system of equations:

$$\frac{EA}{\frac{1}{3}L}\begin{bmatrix} 3 & -3 & 0 & 0 \\ -3 & 5 & -2 & 0 \\ 0 & -2 & 3 & -1 \\ 0 & 0 & -1 & 1 \end{bmatrix}\begin{bmatrix} u_{1X} \\ u_{2X} \\ u_{3X} \\ u_{4X} \end{bmatrix} = \begin{bmatrix} -F_1^{\mathrm{R}} \\ 0 \\ 0 \\ F_0 \end{bmatrix}. \tag{3.126}$$

The reduced system of equations results from taking into account the boundary condition $u_{1X} = 0$ as:

$$\frac{EA}{\frac{1}{3}L}\begin{bmatrix} 5 & -2 & 0 \\ -2 & 3 & -1 \\ 0 & -1 & 1 \end{bmatrix}\begin{bmatrix} u_{2X} \\ u_{3X} \\ u_{4X} \end{bmatrix} = \begin{bmatrix} 0 \\ 0 \\ F_0 \end{bmatrix}. \tag{3.127}$$

The solution is obtained by multiplying the inverse stiffness matrix with the right-hand side:

$$\begin{bmatrix} u_{2X} \\ u_{3X} \\ u_{4X} \end{bmatrix} = \frac{\frac{1}{3}LF_0}{EA}\begin{bmatrix} \frac{1}{3} \\ \frac{5}{6} \\ \frac{11}{6} \end{bmatrix}. \tag{3.128}$$

The highest stress will occur in element III, as it has the smallest cross-sectional area. Based on the nodal displacements, the stress in element III is:

$$\sigma_{\mathrm{III}} = \frac{E}{\frac{L}{3}}(u_{4X} - u_{3X}) = \frac{F_0}{A}. \tag{3.129}$$

The lightweight index is generally defined according to Eq. $(3.1)_{\mathrm{TB}}$ as:

$$M = \frac{F_0}{F_{\mathrm{G}}}, \tag{3.130}$$

where the dead weight is

$$F_{\mathrm{G}} = mg = \varrho g V = \varrho g\left(\frac{3AL}{3} + \frac{2AL}{3} + \frac{AL}{3}\right) = 2\varrho g A L \tag{3.131}$$

and the external load F_0 can be replaced by

$$F_0 = A R_{p0.2} \tag{3.132}$$

using a stress criterion. This ultimately results in the lightweight index as:

$$M = \frac{R_{p0.2}}{2\varrho g L}. \tag{3.133}$$

2.2.11 **Calculation of the lightweight index for a cantilever beam with constant distributed load**

- Internal reactions:

To determine the internal reactions, the cantilever with a constant distributed load is cut free at an arbitrary location x in the inside, see Fig. 3.8. To avoid having to determine the unknown support reactions at the fixed support ($x = 0$), the right-hand section of length $L - x$ is considered.

The moment equilibrium around the reference point at the location x results in the distribution of the internal bending moment:

$$\overset{\curvearrowright}{\sum} M_y = 0: \quad q_0(L-x) \times \frac{(L-x)}{2} - M_y(x) = 0, \tag{3.134}$$

$$M_y(x) = \frac{q_0}{2} \times (L-x)^2. \tag{3.135}$$

The vertical force equilibrium (given here for information only) results in the distribution of the internal shear force:

$$\overset{\downarrow}{\sum} F_z = 0: \quad Q_z(x) + q_0(L-x) = 0, \tag{3.136}$$

$$Q_z(x) = -q_0(L-x). \tag{3.137}$$

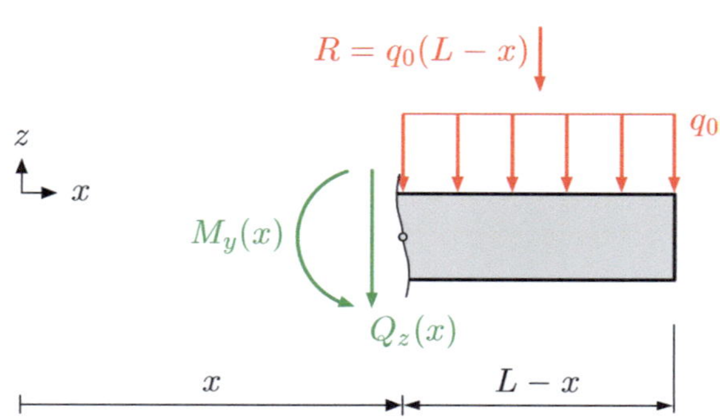

Fig. 3.8 Cutting free inside the cantilever beam with constant distributed load

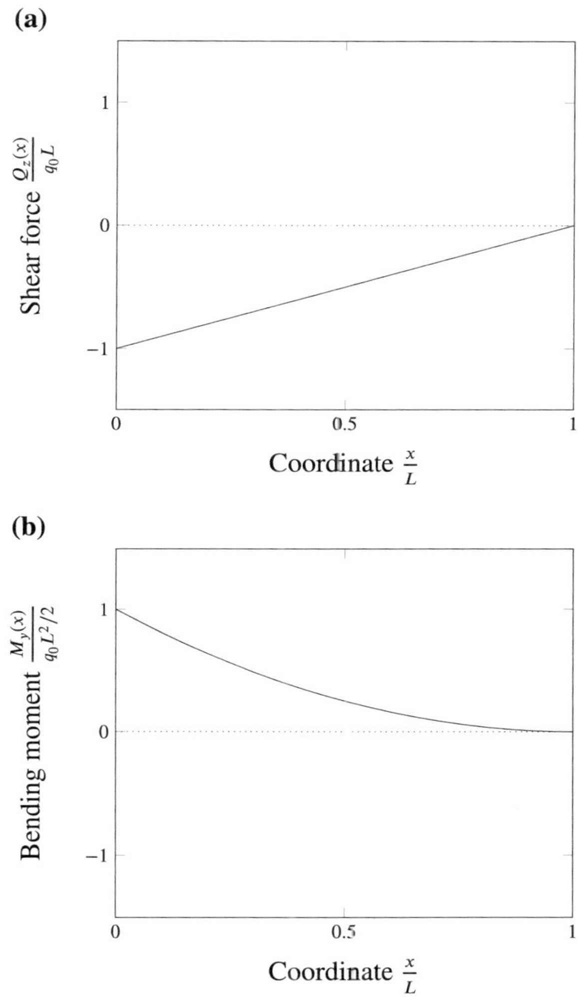

Fig. 3.9 Internal reactions for a cantilever beam with constant distributed load according to Fig. 2.8: **a** shear force and **b** bending moment

The two internal reactions for the problem in Fig. 2.8 are shown in Fig. 3.9.

- Maximum bending moment:

By evaluating the internal bending moment according to Eq. (3.135) at the location $x = 0$ we get:

$$M_{y,\max} = M_y(0) = \frac{q_0 L^2}{2}. \tag{3.138}$$

• Lightweight index:

Since there is no single force in this case, the definition equation of the lightweight index (see Eq. (3.1)$_{TB}$) is evaluated using the resultant of the distributed load:

$$M = \frac{F_0}{F_G} \approx \frac{q_0 L}{F_G}. \tag{3.139}$$

The expression of the resultant ($q_0 L$) is replaced by a stress criterion:

$$\sigma_{max} = \frac{M_{y,max}}{I_y} \times z_{max} = \frac{\frac{q_0 L^2}{2}}{\frac{bh^3}{12}} \times \frac{h}{2} = R_{p0.2}, \tag{3.140}$$

or solved for the resultant:

$$q_0 L = \frac{bh^2 R_{p0.2}}{3L}. \tag{3.141}$$

This ultimately results in the lightweight index:

$$M = \frac{q_0 L}{F_G} = \frac{1}{3} \times \frac{R_{p0.2}}{\varrho g \frac{L^2}{h}}. \tag{3.142}$$

The numerical values of the lightweight index for the various materials (see Table 1.1) are:

$$M_{St} = 1712.015; \; M_{Al} = 4497.575; \; M_{Ti} = 5422.170. \tag{3.143}$$

2.2.12 **Calculation of the lightweight index for a simply supported beam with constant distributed load**

• Internal reactions:

To determine the internal reactions, the simply supported beam, which is subjected to a constant line load, is cut free at an arbitrary location x in the inside, see Fig. 3.10. For example, consider the right-hand section of length $L - x$.

The moment equilibrium around the reference point at the location x results in the distribution of the internal bending moment:

$$\overset{\curvearrowleft}{\sum} M_y = 0: \quad -M_y(x) + q_0(L - x) \times \frac{(L - x)}{2} - \frac{q_0 L}{2} \times (L - x) = 0, \tag{3.144}$$

$$M_y(x) = \frac{q_0}{2} \times (L - x)^2 - \frac{q_0 L}{2} \times (L - x), \tag{3.145}$$

$$= \frac{q_0}{2} \times (x^2 - Lx). \tag{3.146}$$

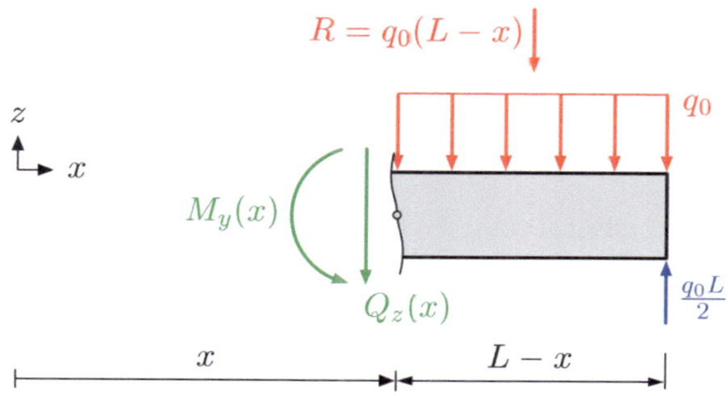

Fig. 3.10 Cutting free inside the simply supported beam with constant distributed load

The vertical force equilibrium (given here for information only) results in the distribution of the internal shear force:

$$\overset{\downarrow}{\sum} F_z = 0: \quad Q_z(x) + q_0(L - x) - \frac{q_0 L}{2} = 0, \tag{3.147}$$

$$Q_z(x) = q_0 x - \frac{q_0 L}{2}. \tag{3.148}$$

The two inner reactions for the problem in Fig. 2.9 are shown in Fig. 3.48.

• Maximum bending moment:

By evaluating the internal bending moment according to Eq. (3.145) at the location $x = L/2$ we get:

$$M_{y,\text{max}} = M_y\left(\tfrac{L}{2}\right) = -\frac{q_0 L^2}{8}. \tag{3.149}$$

• Lightweight index:

Since there is no single force in this case, the definition equation of the lightweight index (see Eq. $(3.1)_{\text{TB}}$) is evaluated using the resultant of the distributed load:

$$M = \frac{F_0}{F_G} \approx \frac{q_0 L}{F_G}. \tag{3.150}$$

The expression of the resultant ($q_0 L$) is replaced by a stress criterion:

$$\sigma_{\text{max}} = \frac{M_{y,\text{max}}}{I_y} \times z_{\text{max}} = \frac{\frac{q_0 L^2}{8}}{\frac{bh^3}{12}} \times \frac{h}{2} = R_{\text{p0.2}}, \tag{3.151}$$

or solved for the resultant:

$$q_0 L = \frac{4bh^2 R_{\text{p0.2}}}{3L}. \tag{3.152}$$

This ultimately results in the lightweight index:

$$M = \frac{q_0 L}{F_{\mathrm{G}}} = \frac{4}{3} \times \frac{R_{\mathrm{p0.2}}}{\varrho g \frac{L^2}{h}}. \tag{3.153}$$

The numerical values of the lightweight index for the various materials (see Table 1.1) are:

$$M_{\mathrm{St}} = 6848.062; \ M_{\mathrm{Al}} = 17990.3001; \ M_{\mathrm{Ti}} = 21688.681. \tag{3.154}$$

2.2.13 **Calculation of the lightweight index for a simply supported beam with point load**

- Internal reactions:

To determine the internal reactions, the simply supported beam, which is loaded by a point load, is cut free at an arbitrary location x in the inside, see Fig. 3.11. For example, consider the left part (i.e., to the left of the acting single force) of length x.

The moment equilibrium around the reference point at the location x results in the distribution of the internal bending moment:

$$\overset{\curvearrowright}{\sum} M_y = 0: \quad +\frac{F_0}{2} \times x + M_y(x) = 0, \tag{3.155}$$

$$M_y(x) = -\frac{F_0}{2} \times x \quad \text{for} \quad 0 \leq x \leq \frac{L}{2}. \tag{3.156}$$

The vertical force equilibrium (given here for information only) results in the distribution of the internal shear force:

$$\overset{\uparrow}{\sum} F_z = 0: \quad +\frac{F_0}{2} + Q_z(x) = 0, \tag{3.157}$$

$$Q_z(x) = -\frac{F_0}{2} \quad \text{for} \quad 0 \leq x \leq \frac{L}{2}. \tag{3.158}$$

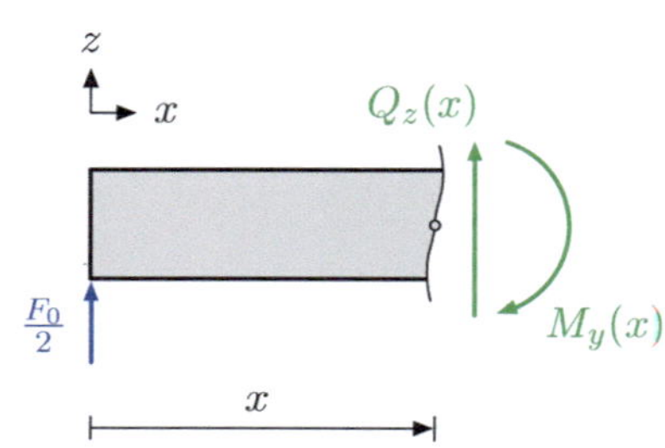

Fig. 3.11 Cutting free inside the simply supported beam with a point load in the middle

The two internal reactions for the problem in Fig. 2.10 are shown in Fig. 3.52.

- Maximum bending moment:

By evaluating the internal bending moment according to Eq. (3.156) at the location $x = L/2$ we get:

$$M_{y,\max} = M_y\left(\tfrac{L}{2}\right) = -\frac{F_0 L}{4}. \tag{3.159}$$

- Lightweight index:

The definition equation of the lightweight index (see Eq. $(3.1)_{\text{TB}}$) results in:

$$M = \frac{F_0}{F_G}. \tag{3.160}$$

The single force (F_0) is replaced by a stress criterion:

$$\sigma_{\max} = \frac{|M_{y,\max}|}{I_y} \times z_{\max} = \frac{\frac{F_0 L}{4}}{\frac{bh^3}{12}} \times \frac{h}{2} = R_{p0.2}, \tag{3.161}$$

or solved for the single force:

$$F_0 = \frac{2bh^2 R_{p0.2}}{3L}. \tag{3.162}$$

This ultimately results in the lightweight index:

$$M = \frac{F_0}{F_G} = \frac{2}{3} \times \frac{R_{p0.2}}{\varrho g \frac{L^2}{h}}. \tag{3.163}$$

The numerical values of the lightweight index for the various materials (see Table 1.1) are:

$$M_{\text{St}} = 3424.031;\ M_{\text{Al}} = 8995.150;\ M_{\text{Ti}} = 10844.340. \tag{3.164}$$

2.2.14 **Calculation of the specific energy absorption for a cantilever beam with constant distributed load**

- Bending moment distribution:

The bending moment distribution can be taken from the solution of Exercise 2.2.11:

$$M_y(x) = \frac{q_0}{2} \times (L - x)^2 = \frac{q_0 L^2}{2}\left(\frac{1}{L^2}x^2 - \frac{2}{L}x + 1\right). \tag{3.165}$$

• Total strain energy:

The total strain energy is calculated according to Eq. (3.32)TB for a bending load case as follows:

$$\Pi = \int_0^L \frac{M_y(x)^2}{2EI_y} \mathrm{d}x. \tag{3.166}$$

Thus, the total strain energy is specifically:

$$\Pi = \frac{1}{2EI_y} \int_0^L \left(\frac{q_0}{2}\right)^2 (L-x)^4 \,\mathrm{d}x = \frac{q_0^2}{8EI_y} \left[\frac{1}{5}(L-x)^5(-1)\right]_0^L = \frac{q_0^2 L^5}{40EI}. \tag{3.167}$$

• Specific energy absorption:

The specific energy absorption is generally given by Eq. (3.27)TB as:

$$SEA = \frac{\Pi}{m}, \tag{3.168}$$

or taking into account the expressions for the total strain energy (see Eq. (3.167)) and mass ($m = \varrho bhL$):

$$SEA = \frac{3q_0^2 L^4}{10b^2h^4 E\varrho}. \tag{3.169}$$

Using a stress criterion, the expression for $q_0 L$ is replaced again, see solution to the Exercise 2.2.11:

$$q_0 L = \frac{bh^2 R_{\mathrm{p0.2}}}{3L}. \tag{3.170}$$

Thus, the specific energy absorption is finally:

$$SEA = \frac{R_{\mathrm{p0.2}}^2}{30E\varrho}. \tag{3.171}$$

2.2.15 **Calculation of the specific energy absorption for a simply supported beam with constant distributed load**

• Bending moment distribution:

The bending moment distribution can be taken from the solution of Exercise 2.2.12:

$$M_y(x) = \frac{q_0}{2}\left(x^2 - Lx\right). \tag{3.172}$$

• Total strain energy:

The total strain energy is calculated according to Eq. (3.32)$_{\text{TB}}$ for a bending load case as follows:

$$\Pi = \int_0^L \frac{M_y(x)^2}{2EI_y} \, dx. \tag{3.173}$$

Thus, the total strain energy is specifically:

$$\Pi = \frac{q_0^2}{8EI_y} \int_0^L (x^4 - 2Lx^3 + L^2x^2) \, dx = \frac{q_0^2}{8EI_y} \left[\frac{x^5}{5} - \frac{Lx^4}{2} + \frac{L^2x^3}{3} \right]_0^L = \frac{q_0^2 L^5}{240EI}. \tag{3.174}$$

• Specific energy absorption:

The specific energy absorption is generally given by Eq. (3.27)$_{\text{TB}}$ as:

$$SEA = \frac{\Pi}{m}, \tag{3.175}$$

or taking into account the expressions for the total strain energy (see Eq. (3.174)) and mass ($m = \varrho bhL$):

$$SEA = \frac{q_0^2 L^4}{20b^2h^4 E\varrho}. \tag{3.176}$$

Using a stress criterion, the expression for $q_0 L$ is replaced again, see solution to the Exercise 2.2.12:

$$q_0 L = \frac{4bh^2 R_{\text{p0.2}}}{3L}. \tag{3.177}$$

Thus, the specific energy absorption is finally:

$$SEA = \frac{4R_{\text{p0.2}}^2}{45E\varrho}. \tag{3.178}$$

2.2.16 **Calculation of the specific energy absorption for a simply supported beam with point load**

• Bending moment distribution:

The bending moment distribution can be taken from the solution of Exercise 2.2.13:

$$M_y(x) = -\frac{F_0 x}{2} \quad \text{for} \quad 0 \le x \le \frac{L}{2}. \tag{3.179}$$

• Total strain energy:

The total strain energy is calculated according to Eq. (3.32)TB for a bending load case as follows:

$$\Pi = \int_0^L \frac{M_y(x)^2}{2EI_y} \mathrm{d}x. \tag{3.180}$$

Due to the symmetry of the problem and the discontinuous distribution of the bending moment, the following formula can be used:

$$\Pi = 2 \times \int_0^{L/2} \frac{M_y(x)^2}{2EI_y} \mathrm{d}x = \frac{1}{EI_y} \int_0^{L/2} M_y(x)^2 \, \mathrm{d}x, \tag{3.181}$$

$$= \frac{F_0^2}{4EI_y} \int_0^{L/2} x^2 = \frac{F_0^2}{4EI_y} \left[\frac{1}{3} x^3 \right]_0^{L/2}, \tag{3.182}$$

$$= \frac{F_0^2 L^3}{96EI} = \frac{F_0^2 L^3}{8Ebh^3}. \tag{3.183}$$

• Specific energy absorption:

The specific energy absorption is generally given by Eq. (3.27)TB as:

$$SEA = \frac{\Pi}{m}, \tag{3.184}$$

or taking into account the expressions for the total strain energy (see Eq. (3.183)) and mass ($m = \varrho bhL$):

$$SEA = \frac{F_0^2 L^2}{8b^2 h^4 E\varrho}. \tag{3.185}$$

Using a stress criterion, the expression for F_0 is replaced again, see solution to the Exercise 2.2.13:

$$F_0 = \frac{2bh^2 R_{\mathrm{p}0.2}}{3L}. \tag{3.186}$$

Thus, the specific energy absorption is finally:

$$SEA = \frac{R_{\mathrm{p}0.2}^2}{18E\varrho}. \tag{3.187}$$

2.2.17 **Calculation of specific energy absorption for a bi-material cantilever beam**

• Bending moment distribution:

The bending moment distribution can be taken from the solution of Exercise 2.1.2 (a):

$$M_y(x) = F_0\,(L - x)\,. \tag{3.188}$$

• Total strain energy:

The total strain energy is calculated according to Eq. (3.32)$_{\text{TB}}$ for a bending load case as follows:

$$\Pi = \int\limits_0^L \frac{M_y(x)^2}{2EI_y}\,\mathrm{d}x. \tag{3.189}$$

Since the beam has two sub-areas with different properties, the integral must be split:

$$\Pi = \int\limits_0^{L/2} \frac{M_y(x)^2}{2E_{\text{I}}I_y}\,\mathrm{d}x + \int\limits_{L/2}^{L} \frac{M_y(x)^2}{2E_{\text{II}}I_y}\,\mathrm{d}x, \tag{3.190}$$

$$= \frac{F_0^2}{2E_{\text{I}}I_y}\int\limits_0^{L/2} (L-x)^2\,\mathrm{d}x + \frac{F_0^2}{2E_{\text{II}}I_y}\int\limits_{L/2}^{L} (L-x)^2\,\mathrm{d}x, \tag{3.191}$$

$$= \frac{F_0^2}{2E_{\text{I}}I_y}\left[\frac{1}{3}\,(L-x)^3(-1)\right]_0^{L/2} + \frac{F_0^2}{2E_{\text{II}}I_y}\left[\frac{1}{3}\,(L-x)^3(-1)\right]_{L/2}^{L}, \tag{3.192}$$

$$= \frac{F_0^2L^3}{2E_{\text{I}}I_y}\times\frac{7}{24} + \frac{F_0^2L^3}{2E_{\text{II}}I_y}\times\frac{1}{24}, \tag{3.193}$$

$$= \frac{F_0^2L^3}{48I_y}\times\left(\frac{7}{E_{\text{I}}} + \frac{1}{E_{\text{II}}}\right). \tag{3.194}$$

• Mass:

The total mass is the sum of the two sub-areas:

$$m = m_{\text{I}} + m_{\text{II}} = \frac{a^2L}{2}\times(\varrho_{\text{I}} + \varrho_{\text{II}})\,. \tag{3.195}$$

• Specific energy absorption:

The specific energy absorption is generally given by Eq. (3.27)$_{\text{TB}}$ as:

$$SEA = \frac{\Pi}{m}, \tag{3.196}$$

or taking into account the expressions for the total strain energy (see Eq. (3.194)) and mass (see Eq. (3.195)):

$$SEA = \frac{F_0^2 L^2}{2a^6} \times \frac{\frac{7}{E_{\mathrm{I}}} + \frac{1}{E_{\mathrm{II}}}}{\varrho_{\mathrm{I}} + \varrho_{\mathrm{II}}}. \tag{3.197}$$

3.3 Lightweight Design Concepts: Shape Optimization

2.3.18 **Analysis of the lightweight index for a cantilever beam with end force and tubular cross section**

• Bending moment distribution:

The bending moment distribution can be taken from the solution of Exercise 2.1.2 (a):

$$M_y(x) = F_0\,(L - x)\,. \tag{3.198}$$

This results in the maximum bending moment at the clamping point ($x = 0$) being:

$$M_{y,\mathrm{max}} = M_y(x = 0) = F_0 L. \tag{3.199}$$

• Geometric properties of the cross section:

The expressions for the axial second moment of area and the cross-sectional area can generally be taken from the solution of Exercise 2.1.5 for a circular cross section with diameter d (or radius r):

$$I_y = \frac{\pi}{4}\,r^4 = \frac{\pi}{64}\,d^4, \tag{3.200}$$

$$A = \pi r^2. \tag{3.201}$$

With the dimensions given in this exercise, i.e. outer diameter d and wall thickness s, we get:

$$I_y = \frac{\pi}{64}\left(d^4 - (d - 2\,s)^4\right), \tag{3.202}$$

$$= \frac{\pi}{64}\left(d^4 - (16\,s^4 - 32 d s^3 + 24 d^2 s^2 - 8 d^3 s + d^4)\right), \tag{3.203}$$

$$= \frac{\pi}{16}\left(-4\,s^4 + 8 d s^3 - 6 d^2 s^2 + 2 d^3 s\right), \tag{3.204}$$

or for the cross-sectional area:

$$A = \pi\left(\left(\frac{d}{2}\right)^2 - \left(\frac{d}{2} - s\right)^2\right), \tag{3.205}$$

$$= \pi\left(d s - s^2\right). \tag{3.206}$$

• Lightweight index:

The definition equation of the lightweight index (see Eq. $(3.1)_{TB}$) results in:

$$M = \frac{F_0}{F_G}. \tag{3.207}$$

The single force (F_0) is replaced by a stress criterion:

$$\sigma_{\max} = \frac{|M_{y,\max}|}{I_y} \times z_{\max} = \frac{F_0 L}{I_y} \times \frac{d}{2} = R_{p0.2}, \tag{3.208}$$

or solved for the single force:

$$F_0 = \frac{2 R_{p0.2} I_y}{dL}. \tag{3.209}$$

With this relationship for the force F_0 and the geometric dimensions, the lightweight index is finally:

$$M(d,s) = \frac{F_0(d,s)}{F_G(d,s)} = \frac{R_{p0.2}}{4\varrho g L^2} \times \frac{\frac{1}{d}\left(-2s^4 + d^3 s + 4ds^3 - 3d^2 s^2\right)}{ds - s^2} \tag{3.210}$$

$$= \frac{R_{p0.2}}{4\varrho g L^2} \times \frac{F_0^*(d,s)}{F_G^*(d,s)} \tag{3.211}$$

$$= \frac{R_{p0.2}}{4\varrho g L^2} \times \frac{-2s^3 + d^3 + 4ds^2 - 3d^2 s}{d(d-s)}. \tag{3.212}$$

The functions $F_0(d,s)$, $F_G(d,s)$ and $M(d,s) = F_0(d,s)/F_G(d,s)$ are shown in Fig. 3.12.

2.3.19 **Implementation of the concept of lightweight design shapes: Optimization of a box profile**

• Geometric condition $A^{\boxtimes} = A^{\square}$:

The equality of the cross-sectional areas results in the following relationship:

$$h^2 = ab - (b - 2w)(a - 2w), \tag{3.213}$$

$$\overset{w=h/5}{=} ab - \left(b - \frac{2h}{5}\right)\left(a - \frac{2h}{5}\right), \tag{3.214}$$

$$= \frac{2}{5}bh + \frac{2}{5}ah - \frac{4}{25}h^2, \tag{3.215}$$

$$\rightarrow h = \frac{2}{5}b + \frac{2}{5}a - \frac{4}{25}h, \tag{3.216}$$

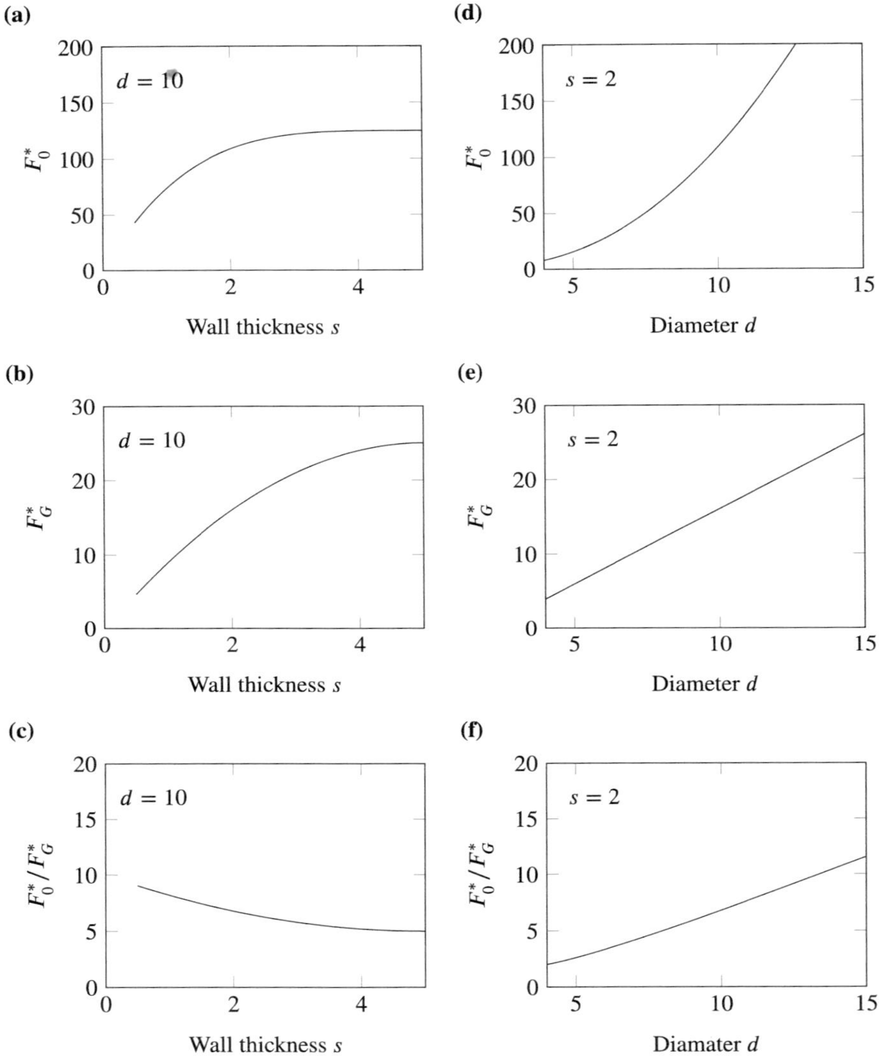

Fig. 3.12 Functions $F_0(d, s)$, $F_G(d, s)$ and $M(d, s) = F_0(d, s)/F_G(d, s)$ depending on s at $d = 10\,\text{mm} = \text{const.}$ ((**a**)–(**c**)) and from d at $s = 2\,\text{mm} = \text{const.}$ ((**d**)–(**f**))

respectively

$$b = \frac{29}{10}h - a, \tag{3.217}$$

or also as:

$$\frac{a}{h} = \frac{29}{10}\left(\frac{b}{a} + 1\right)^{-1}, \tag{3.218}$$

or resolved according to the width-to-height ratio:

$$\frac{a}{b} = \frac{10}{29}\left(\frac{h}{a} - 1\right)^{-1} = \frac{10}{29}\left(\left(\frac{a}{h}\right)^{-1} - 1\right)^{-1}. \tag{3.219}$$

• Second moment of area of the box profile:

$$I_y^{\square} = \frac{b\,a^3}{12} - \frac{1}{12}\,(b - 2w)\,(a - 2w)^3, \tag{3.220}$$

$$= -\frac{4w^4}{3} + \frac{2b\,w^3}{3} + 2a\,w^3 - ab\,w^2 - a^2\,w^2 + \frac{a^2bw}{2} + \frac{a^3w}{6}, \tag{3.221}$$

or by means of $w = \frac{h}{5}$

$$I_y^{\square} = -\frac{4h^4}{1875} + \frac{2b\,h^3}{375} + \frac{2a\,h^3}{125} - \frac{ab\,h^2}{25} - \frac{a^2\,h^2}{25} + \frac{a^2bh}{10} + \frac{a^3h}{30}, \tag{3.222}$$

or by means of $b = \frac{29h}{10} - a$ (see Eq. (3.217))

$$I_y^{\square} = \frac{h^4}{75} - \frac{79a\,h^3}{750} + \frac{29a^2\,h^2}{100} - \frac{a^3h}{15}, \tag{3.223}$$

or in normalized representation:

$$\frac{I_y^{\square}}{h^4} = \frac{1}{75} - \frac{79}{750}\left(\frac{a}{h}\right) + \frac{29}{100}\left(\frac{a}{h}\right)^2 - \frac{1}{15}\left(\frac{a}{h}\right)^3. \tag{3.224}$$

The course of the second moment of area is shown in Fig. 3.13.

• Lightweight index:

The definition equation of the lightweight index (see Eq. $(3.1)_{\text{TB}}$) results in:

$$M = \frac{F_0}{F_{\text{G}}}. \tag{3.225}$$

The single force (F_0) is replaced by a stress criterion:

$$\sigma_{\max} = \frac{|M_{y,\max}|}{I_y} \times z_{\max} = \frac{F_0 L}{I_y} \times \frac{a}{2} = R_{\text{p}0.2}, \tag{3.226}$$

or solved for the single force:

$$F_0 = \frac{2I_y\,R_{\text{p}0.2}}{aL}. \tag{3.227}$$

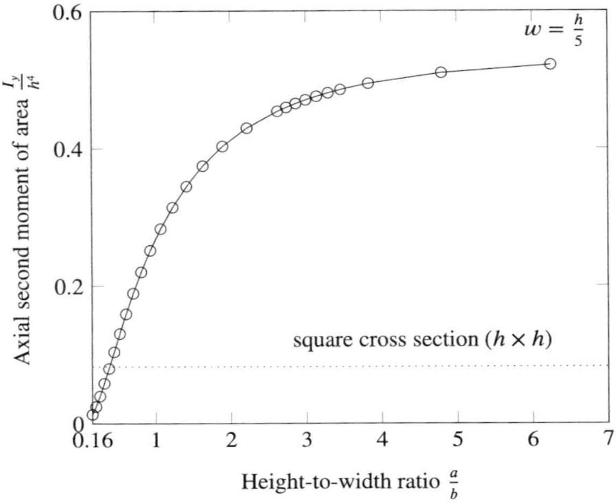

Fig. 3.13 Axial second moment of area for different box profiles. Same cross-sectional area ($A^{\boxtimes} = A^{\square}$) and thus same mass ($m^{\boxtimes} = m^{\square}$) assumed

This ultimately results in the lightweight index:

$$M == \frac{R_{\text{St}}}{\varrho_{\text{St}} g L^2 h^2} \times \frac{I_y^{\square}}{\frac{a}{2}} \tag{3.228}$$

$$= \frac{R_{\text{St}}}{\varrho_{\text{St}} g L^2 h^2} \times \frac{\frac{h^4}{75} - \frac{79ah^3}{750} + \frac{29a^2h^2}{100} - \frac{a^3h}{15}}{\frac{a}{2}} \tag{3.229}$$

$$= \frac{R_{\text{St}}}{\varrho_{\text{St}} g L^2 h^2} \times \frac{\left(\frac{I_y^{\square}}{h^4}\right) \times h^4}{\frac{1}{2}\left(\frac{a}{h}\right) \times h}. \tag{3.230}$$

The course of the lightweight index is shown in Fig. 3.14.

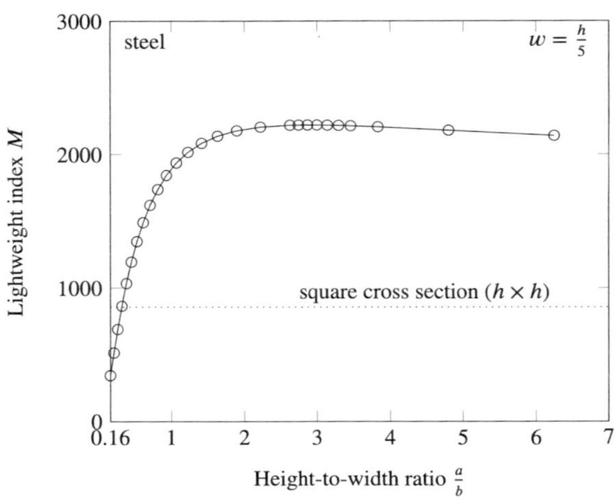

Fig. 3.14 Lightweight index for different box profiles and stress criterion ($\sigma_{\max} = R_{p0.2}$). Same cross-sectional area ($A^{\boxtimes} = A^{\square}$) and thus same mass ($m^{\boxtimes} = m^{\square}$) assumed

The graphical representation shown in Figs. 3.13 and 3.14 was created using the following procedure: The markers in both figures correspond to a calculated coordinate in the respective coordinate system, which was then approximated using a best-fit curve. By specifying numerical values of $\frac{a}{h}$, the width-to-height ratio $\frac{a}{b}$ can be calculated using Eq. (3.219). The normalized axial second moment of area is obtained for a given $\frac{a}{h}$ according to Eq. (3.224), and the lightweight index can then be calculated using Eq. (3.230).

The pairwise assignment for creating the diagrams is summarized in Table 3.2.

Finally, it should be noted that the highest lightweight index of $M_{\max} = 2220.84$ is achieved for a width-to-height ratio of $\frac{a}{b} = 2.867$.

Table 3.2 Pairs of values for creating Figs. 3.13 and 3.14

$\frac{a}{h}$	$\frac{a}{b}$	$\frac{I_y}{h^4}$	M
0.4	0.16	0.013333	342.403095
0.5	0.2083333	0.024833	510.180611
0.6	0.260870	0.040133	687.088877
0.7	0.318182	0.058833	863.344946
0.8	0.380952	0.080533	1034.057346
0.9	0.45	0.104833	1196.508592
1.0	0.526316	0.131333	1349.068193
1.1	0.611111	0.159633	1490.698564
1.2	0.705882	0.189333	1620.707982
1.3	0.8125	0.220033	1738.617560
1.4	0.933333	0.251333	1844.085239
1.5	1.071429	0.2828333	1936.860172
1.6	1.230769	0.314133	2016.754228
1.7	1.416667	0.344833	2083.623538
1.8	1.636364	0.374533	2137.356207
1.9	1.9	0.402833	2177.863894
2.0	2.222222	0.429333	2205.075930
2.1	2.625	0.453633	2218.935103
2.125	2.741935	0.459318	2220.307832
2.15	2.866667	0.464833	2220.842398
2.175	3.0	0.470174	2220.538185
2.2	3.142857	0.475333	2219.394605
2.225	3.296296	0.480305	2217.411098
2.25	3.461538	0.485083	2214.587127
2.3	3.833333	0.494033	2206.415768
2.4	4.8	0.509333	2179.966370
2.5	6.25	0.520833	2140.019342

2.3.20 **Optimization of a beam in bending along its axis**

The bending moment distribution can be taken from the solution of Exercise 2.1.2 (a):

$$M_y(x) = F_0 L \left(1 - \frac{x}{L}\right) = F_0\,(L - x)\,. \tag{3.231}$$

This results in the maximum bending moment at the clamping point ($x = 0$) being:

$$M_{y,\text{max}} = M_y(x = 0) = F_0 L. \tag{3.232}$$

• Reference configuration:

The cross-sectional area of the reference configuration is:

$$\begin{aligned} A_0^{\circledcirc} &= \frac{\pi}{4}\left(d_0^2 - d_\text{i}^2\right) = \frac{\pi}{4}\left(d_0^2 - (d_0 - 2s_0)^2\right) = \frac{\pi}{4}\left(d_0^2 - (d_0 - \tfrac{2d_0}{5})^2\right) \\ &= \frac{4\pi}{25} d_0^2. \end{aligned} \tag{3.233}$$

Alternatively, the cross-sectional area of the reference configuration can be determined from the solution to the Exercise 2.1.5:

$$A = \pi\,(r_\text{o} + r_\text{i})\,s = \pi\left(\frac{d_\text{o}}{2} + \frac{d_\text{o}}{2} - s\right)s = \pi(d_\text{o} - s)s. \tag{3.234}$$

With $d_\text{o} \to d_0$ and $s \to s_0 = d_0/5$, we get Eq. (3.233). The axial second moment of area of the reference configuration is:

$$\begin{aligned} I_{y,0}^{\circledcirc} &= \frac{\pi}{64}\left(d_0^4 - d_\text{i}^4\right) = \frac{\pi}{64}\left(d_0^4 - (d_0 - 2s_0)^4\right) = \frac{\pi}{64}\left(d_0^4 - (d_0 - \tfrac{2d_0}{5})^4\right) \\ &= \frac{17}{1250}\pi d_0^4. \end{aligned} \tag{3.235}$$

Alternatively, the axial second moment of area of the reference configuration can be determined from the solution to the Exercise 2.1.5:

$$I_y = \frac{\pi}{8} d_\text{m}^3\, s \left(1 + \frac{s^2}{d_\text{m}^2}\right). \tag{3.236}$$

With $d_\text{m} = \frac{1}{2}(d_\text{o} + d_\text{i}) = d_\text{a} - s$ this gives:

$$I_y = \frac{\pi}{8}(d_\text{o} - 3)^3\, s \left(1 + \frac{s^2}{(d_\text{o} - s)^2}\right). \tag{3.237}$$

With $d_o \rightarrow d_0$ and $s \rightarrow s_0 = d_0/5$, we get Eq. (3.235). The mass of the reference configuration is:

$$m_0^{\circledcirc} = \varrho V_0^{\circledcirc} = \varrho A_0^{\circledcirc} L = \varrho \frac{4\pi}{25} d_0^2 L. \tag{3.238}$$

Using the stress criterion

$$\sigma_{x,\max} = \frac{M_{y,\max}}{I_{y,0}^{\circledcirc}} \times \frac{d_0}{2} \stackrel{!}{=} R_{p0.2}, \tag{3.239}$$

the maximum value of the force is obtained as:

$$F_0 = \frac{R_{p0.2} I_{y,0}^{\circledcirc}}{\frac{1}{2} d_0 L}. \tag{3.240}$$

The definition equation of the lightweight index (see Eq. $(3.1)_{\text{TB}}$) results in:

$$M = \frac{F_0}{F_G}. \tag{3.241}$$

From this, using Eqs. (3.240) and (3.238), the lightweight index of the reference configuration is:

$$M_0 = \frac{F_0}{F_G} = \frac{17}{100} \times \frac{R_{p0.2}}{\varrho g \frac{L^2}{d_0}}. \tag{3.242}$$

The total strain energy is calculated according to Eq. $(3.32)_{\text{TB}}$ for a bending load case as follows:

$$\Pi = \int_0^L \frac{M_y(x)^2}{2EI_y} \mathrm{d}x. \tag{3.243}$$

For the reference configuration in our case, this results in:

$$\Pi_0 = \int_0^L \frac{M_y(x)^2}{2EI_{y,0}^{\circledcirc}} \mathrm{d}x = \frac{F_0^2}{2EI_{y,0}^{\circledcirc}} \int_0^L \left(L^2 - 2Lx + x^2\right) \mathrm{d}x \tag{3.244}$$

$$= \frac{F_0^2}{2EI_{y,0}^{\circledcirc}} \left[L^2 x - Lx^2 + \frac{1}{3} x^3 \right]_0^L \tag{3.245}$$

$$= \frac{F_0^2 L^3}{6EI_{y,0}^{\circledcirc}}. \tag{3.246}$$

Using Eq. (3.240) we get:

$$\Pi_0 = \frac{L^3}{6EI_{y,0}^{\circledcirc}} \times \frac{R_{p0.2}^2 I_{y,0}^{\circledcirc,2}}{\frac{1}{4}d_0^2 L^2} = \frac{L R_{p0.2}^2 I_{y,0}^{\circledcirc}}{\frac{3}{2}E d_0^2}. \tag{3.247}$$

The specific energy absorption is generally given by Eq. (3.27)TB as:

$$SEA = \frac{\Pi}{m}. \tag{3.248}$$

For the reference configuration in our case, this results in:

$$SEA_0 = \frac{\Pi_0}{m_0} = \frac{L R_{p0.2}^2 I_{y,0}^{\circledcirc}}{\frac{3}{2}E d_0^2} \times \frac{1}{\varrho \frac{4\pi}{25} d_0^2 L} = \frac{17}{300} \times \frac{R_{p0.2}^2}{E\varrho}. \tag{3.249}$$

- Optimized configuration:

Using the stress criterion

$$\sigma_{x,\max} = \frac{M_y(x)}{I_y^{\circledcirc}(x)} \times \frac{d(x)}{2} = \frac{F_0 L (1-\frac{x}{L})}{\frac{17}{1250}\pi d^4(x)} \times \frac{d(x)}{2} \stackrel{!}{=} R_{p0.2}, \tag{3.250}$$

we obtain

$$d^3(x) = \frac{F_0 L}{\frac{17}{1250}\pi 2 R_{p0.2}} \times \left(1 - \tfrac{x}{L}\right). \tag{3.251}$$

By evaluating the stress criterion at $x = 0$, i.e., for the reference configuration, we get:

$$\sigma_{x,\max} = \frac{M_y(x=0)}{I_{y,0}^{\circledcirc}(x)} \times \frac{d_0}{2} = \frac{F_0 L}{\frac{17}{1250}\pi d_0^4} \times \frac{d_0}{2} \stackrel{!}{=} R_{p0.2}, \tag{3.252}$$

or resolved according to the reference diameter:

$$d_0^3 = \frac{F_0 L}{\frac{17}{1250}\pi 2 R_{p0.2}}. \tag{3.253}$$

This results from Eq. (3.251) the following expression:

$$d^3(x) = d_0^3 \left(1 - \tfrac{x}{L}\right). \tag{3.254}$$

Thus, the outer diameter of the beam cross section is (see Fig. 3.15):

$$d(x) = d_0 \times \sqrt[3]{1 - \frac{x}{L}}. \tag{3.255}$$

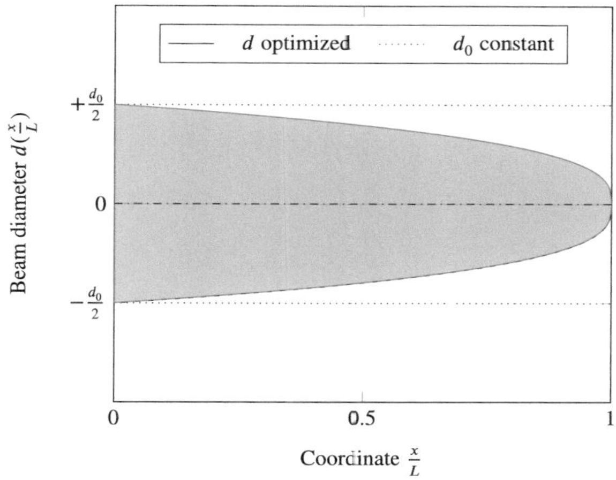

Fig. 3.15 Beam diameter (outer) along the beam axis

The volume of the optimized configuration must be calculated by integration, since the cross section changes continuously with the x-coordinate:

$$V = \int_0^L A(x)\mathrm{d}x = \int_0^L \frac{4\pi}{25} d^2(x)\mathrm{d}x = \frac{4\pi d_0^2}{25} \int_0^L \left(1 - \frac{x}{L}\right)^{\frac{2}{3}} \mathrm{d}x \tag{3.256}$$

$$= \frac{4\pi d_0^2}{25} \left[\frac{3}{5} \left(1 - \frac{x}{L}\right)^{\frac{5}{3}} (-L) \right]_0^L \tag{3.257}$$

$$= \frac{12\pi}{125} d_0^2 L. \tag{3.258}$$

With this volume, the mass of the optimized configuration is:

$$m = \varrho V = \frac{12\pi}{125} d_0^2 L \varrho. \tag{3.259}$$

Thus, the lightweight index of the optimized configuration is (the force F_0 can be further replaced using Eq. (3.240)):

$$M = \frac{F_0}{F_G} = \frac{R_{p0.2} I_{y,0}^{\circledcirc}}{\frac{1}{2} d_0 L} \times \frac{1}{\varrho g \frac{12\pi}{125} d_0^2 L} = \frac{17}{60} \times \frac{R_{p0.2}}{\varrho g \frac{L^2}{d_0}}. \tag{3.260}$$

The total strain energy of the optimized configuration is calculated as:

$$\Pi = \int_0^L \frac{M_y(x)^2}{2EI_y^{\circledcirc}(x)}\mathrm{d}x = \frac{F_0^2L^2}{2EI_{y,0}^{\circledcirc}}\int_0^L \frac{(1-\frac{x}{L})^2}{(1-\frac{x}{L})^{\frac{4}{3}}}\mathrm{d}x = \frac{F_0^2L^2}{2EI_{y,0}^{\circledcirc}}\int_0^L (1-\tfrac{x}{L})^{\frac{2}{3}}\,\mathrm{d}x$$

$$= \frac{F_0^2L^2}{2EI_{y,0}^{\circledcirc}}\left[\frac{3}{5}\left(1-\frac{x}{L}\right)^{\frac{5}{3}}(-L)\right]_0^L \tag{3.261}$$

$$= \frac{3}{10}\times\frac{F_0^2L^3}{EI_{y,0}^{\circledcirc}}. \tag{3.262}$$

Using Eq. (3.240) this gives:

$$\Pi = \frac{3}{10}\frac{L^3}{EI_{y,0}^{\circledcirc}}\times\frac{R_{\mathrm{p}0.2}^2 I_{y,0}^{\circledcirc,2}}{\frac{1}{4}d_0^2L^2} = \frac{6}{5}\times\frac{LR_{\mathrm{p}0.2}^2 I_{y,0}^{\circledcirc}}{Ed_0^2}, \tag{3.263}$$

respectively for the specific energy absorption of the optimized configuration:

$$SEA = \frac{\Pi}{m} = \frac{6}{5}\times\frac{LR_{\mathrm{p}0.2}^2 I_{y,0}^{\circledcirc}}{Ed_0^2}\times\frac{1}{\frac{12\pi}{125}d_0^2L\varrho} = \frac{17}{100}\times\frac{R_{\mathrm{p}0.2}^2}{E\varrho}. \tag{3.264}$$

- Ratios:

$$\frac{m}{m_0} = \frac{3}{5} \approx 0.6, \tag{3.265}$$

$$\frac{M}{M_0} = \frac{5}{3} \approx 1.667, \tag{3.266}$$

$$\frac{SEA}{SEA_0} = 3.0. \tag{3.267}$$

2.3.21 **Optimization of a cantilevered beam loaded with two single forces of the same magnitude**

- Internal reactions:

To determine the internal reactions, the beam under the influence of two individual forces is cut free at an arbitrary location x to the left and then to the right of the left single force, see Fig. 3.16.

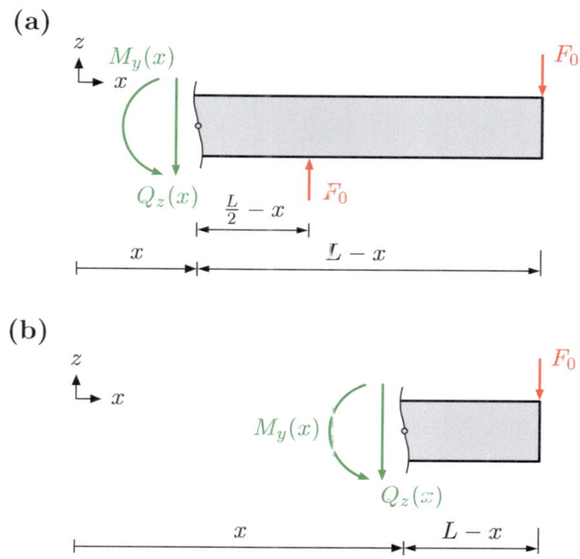

Fig. 3.16 Cutting free inside the beam under the influence of two individual forces: **a** Section 1 and **b** Section 2

The moment equilibrium around the reference point at the location x in the range 1 ($0 \leq x \leq \frac{L}{2}$) results in the distribution of the internal bending moment, see Fig. 3.16a:

$$\overset{\curvearrowright}{\sum} M_y = 0: \quad +F_0(L - x) - F_0\left(\tfrac{L}{2} - x\right) - M_y(x) = 0, \tag{3.268}$$

$$M_y(x) = \frac{F_0 L}{2} \quad \text{for} \quad 0 \leq x \leq \frac{L}{2}. \tag{3.269}$$

The moment equilibrium around the reference point at the location x in the range 2 ($\frac{L}{2} \leq x \leq L$) results in the distribution of the internal bending moment, see Fig. 3.16b:

$$\overset{\curvearrowright}{\sum} M_y = 0: \quad +F_0(L - x) - M_y(x) = 0, \tag{3.270}$$

$$M_y(x) = F_0(L - x) \quad \text{for} \quad \frac{L}{2} \leq x \leq L. \tag{3.271}$$

• Optimized diameter:

Using the stress criterion in the range 1 ($0 \leq x \leq \frac{L}{2}$)

$$\sigma_{x,\max} = \frac{M_y(x)}{I_y^{\circledcirc}(x)} \times \frac{d(x)}{2} = \frac{\frac{F_0 L}{2}}{\frac{17}{1250}\pi d^4(x)} \times \frac{d(x)}{2} \overset{!}{=} R_{p0.2}, \tag{3.272}$$

results (the evaluation is also valid at the location $x = 0$)

$$d^3(x) = \frac{625 F_0 L}{34\pi R_{p0.2}} = d_0^3. \tag{3.273}$$

Using the stress criterion in the range 2 ($\frac{L}{2} \leq x \leq L$)

$$\sigma_{x,\max} = \frac{M_y(x)}{I_y^{\circledcirc}(x)} \times \frac{d(x)}{2} = \frac{F_0(L-x)}{\frac{17}{1250}\pi d^4(x)} \times \frac{d(x)}{2} \stackrel{!}{=} R_{p0.2}, \tag{3.274}$$

results

$$d^3(x) = \frac{625 F_0(L-x)}{17\pi R_{p0.2}} = 2 \times d_0^3 \left(1 - \frac{x}{L}\right). \tag{3.275}$$

In summary, the following distribution results for both sections:

$$d(x) = d_0 \quad \text{for} \quad 0 \leq x \leq \frac{L}{2}, \tag{3.276}$$

$$d(x) = d_0 \left(2 - \frac{2x}{L}\right)^{\frac{1}{3}} \quad \text{for} \quad \frac{L}{2} \leq x \leq L, \tag{3.277}$$

or as a *single* equation using the Föppl-symbol[2] (see Fig. 3.17):

$$d(x) = d_0 \left(1 + \left(1 - \frac{2x}{L}\right)\left\langle x - \frac{L}{2}\right\rangle^0\right)^{\frac{1}{3}}. \tag{3.278}$$

- Axial second moment of area:

For a tube with outer diameter d and wall thickness $s = d/5$, the relationship for the axial second moment of area can be taken from the solution of Exercise 2.3.20. This results in the following for both sections:

$$I_y^{\circledcirc} = \frac{17}{1250} \times \pi d_0^4 \quad \text{for} \quad 0 \leq x \leq \frac{L}{2}, \tag{3.279}$$

$$I_y^{\circledcirc} = \frac{17}{1250} \times \pi d_0^4 (16)^{\frac{1}{3}} \left(1 - \frac{x}{L}\right)^{\frac{4}{3}} \quad \text{for} \quad \frac{L}{2} \leq x \leq L, \tag{3.280}$$

[2] In the English literature, this is also related to the British mathematician William Herrick Macaulay.

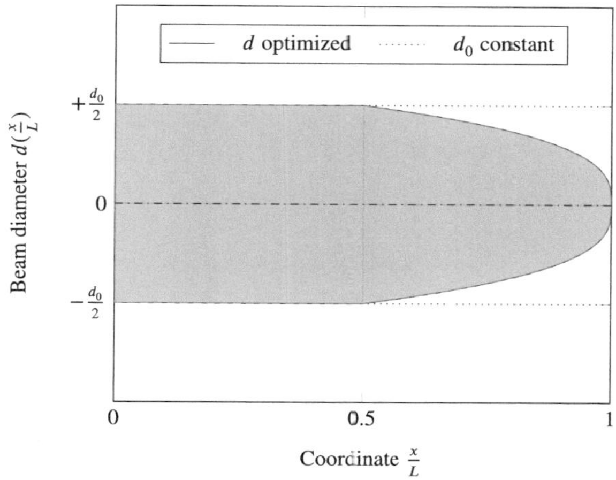

Fig. 3.17 Beam diameter (outer) along the beam axis

or as a *single* equation using the Föppl-symbol:

$$I_y^{\circledcirc} = \frac{17}{1250} \times \pi d_0^4 \left(1 + \left(1 - \frac{2x}{L}\right)\left\langle x - \frac{L}{2}\right\rangle^0\right)^{\frac{4}{3}}. \tag{3.281}$$

• Total strain energy:

$$\Pi = \int_0^{\frac{L}{2}} \frac{M_y^2}{2EI_y^{\circledcirc}} \mathrm{d}x + \int_{\frac{L}{2}}^{L} \frac{M_y(x)^2}{2EI_y^{\circledcirc}(x)} \mathrm{d}x \tag{3.282}$$

$$= \frac{\frac{1}{4} F_0^2 L^2}{2EI_y^{\circledcirc}} \int_0^{\frac{L}{2}} 1\,\mathrm{d}x + \frac{F_0^2}{2E\frac{17}{1250}\pi d_0^4 (16)^{\frac{1}{3}}} \int_{\frac{L}{2}}^{L} \frac{L^2(1-\frac{x}{L})^2}{(1-\frac{x}{L})^{\frac{4}{3}}} \mathrm{d}x \tag{3.283}$$

$$= \frac{F_0^2 L^3}{16EI_y^{\circledcirc}} + \frac{F_0^2 L^2}{2EI_y^{\circledcirc}(16)^{\frac{1}{3}}} \int_{\frac{L}{2}}^{L} \left(1 - \frac{x}{L}\right)^{\frac{2}{3}} \mathrm{d}x \tag{3.284}$$

$$= \frac{F_0^2 L^3}{16EI_y^{\circledcirc}} + \frac{F_0^2 L^2}{2EI_y^{\circledcirc}(16)^{\frac{1}{3}}} \left[\frac{3}{5}\left(1 - \frac{x}{L}\right)^{\frac{5}{3}} (-L)\right]_{\frac{L}{2}}^{L} \tag{3.285}$$

$$= \frac{125}{17} \times \frac{F_0^2 L^3}{\pi d_0^4 E}. \tag{3.286}$$

- Volume:

For a tube with outer diameter d and wall thickness $s = d/5$, the relationship for the cross-sectional area can be taken from the solution to Exercise 2.3.20. Thus, the total volume is:

$$V = \int_0^{\frac{L}{2}} A_0 \,\mathrm{d}x + \int_{\frac{L}{2}}^{L} A(x)\,\mathrm{d}x \tag{3.287}$$

$$= \int_0^{\frac{L}{2}} \frac{4\pi}{25} d_0^2 \,\mathrm{d}x + \int_{\frac{L}{2}}^{L} \frac{4\pi}{25} d_0^2 2^{\frac{2}{3}} \left(1 - \frac{x}{L}\right)^{\frac{2}{3}} \mathrm{d}x \tag{3.288}$$

$$= \frac{16}{125} \times \pi d_0^2 L. \tag{3.289}$$

- Specific energy absorption:

The specific energy absorption is generally given by Eq. $(3.27)_{\mathrm{TB}}$ as:

$$SEA = \frac{\Pi}{m}. \tag{3.290}$$

The expression for the force F_0 in the total strain energy according to Eq. (3.286) can be replaced using the stress criterion at $x = 0$:

$$\sigma_{x,\max} = \frac{M_y(x)}{I_y^{\circledcirc}(x)} \times \frac{d(x)}{2} = \frac{\frac{F_0 L}{2}}{\frac{17}{1250}\pi d_0^4} \times \frac{d_0}{2} \stackrel{!}{=} R_{\mathrm{p}0.2}, \tag{3.291}$$

or solved for the force:

$$F_0 = \frac{34}{625} \times \frac{\pi d_0^3 R_{\mathrm{p}0.2}}{L}. \tag{3.292}$$

Thus, the specific energy absorption is:

$$SEA = \frac{\Pi}{m} = \frac{125^2}{16 \times 17} \times \frac{F_0^2 L^2}{\pi^2 d_0^6 \varrho E} = \frac{17}{100} \times \frac{R_{\mathrm{p}0.2}^2}{\varrho E}. \tag{3.293}$$

2.3.22 **Optimization of a beam under 3-point bending: tubular cross section**

• Internal reactions:

The moment distribution can be taken from the solution of the Exercise 2.2.13:

$$M_y(x) = -\frac{F_0}{2} \times x \quad \text{for} \quad 0 \leq x \leq \frac{L}{2}. \tag{3.294}$$

Due to the symmetry of the problem, the following moment distribution results for the right part of the beam:

$$M_y(x) = -\frac{F_0}{2} \times (L - x) \quad \text{for} \quad \frac{L}{2} \leq x \leq L, \tag{3.295}$$

where the maximum value of the bending moment is in the middle of the beam:

$$M_{y,\text{max}} = M_y\left(x = \tfrac{L}{2}\right) = -\frac{F_0 L}{4}. \tag{3.296}$$

• Axial second moment of area and cross-sectional area:

For a tube with outer diameter d_0 and wall thickness $s_0 = d_0/5$, the relationship for the axial second moment of area and the cross-sectional area can be taken from the solution of Exercise 2.3.20:

$$I_{y,0}^{\circledcirc} = \frac{17}{1250} \pi d_0^4. \tag{3.297}$$

$$A_0^{\circledcirc} = \frac{4\pi}{25} d_0^2. \tag{3.298}$$

• Optimized diameter:

Using the stress criterion in the range 1 ($0 \leq x \leq \frac{L}{2}$)

$$\sigma_{x,\text{max}} = \frac{M_y(x)}{I_y^{\circledcirc}(x)} \times \frac{d(x)}{2} = \frac{\frac{F_0 x}{2}}{\frac{17}{1250}\pi d^4(x)} \times \frac{d(x)}{2} \stackrel{!}{=} R_{\text{p0.2}}, \tag{3.299}$$

results

$$d^3(x) = \frac{1250 F_0 L}{4 \times 17\pi R_{\text{p0.2}}} \times x. \tag{3.300}$$

The evaluation of the stress criterion in the reference cross section in the middle of the beam at $x = \frac{L}{2}$, i.e.

$$\sigma_{x,\text{max}} = \frac{M_{y,\text{max}}}{I_y^{\circledcirc}} \times \frac{d_0}{2} = \frac{\frac{F_0 L}{4}}{\frac{17}{1250}\pi d_0^4} \times \frac{d_0}{2} \stackrel{!}{=} R_{\text{p0.2}}, \tag{3.301}$$

gives the reference diameter as:

$$d_0 = \left(\frac{F_0 L}{2 \times \frac{68}{1250} \pi R_{p0.2}} \right)^{\frac{1}{3}} . \tag{3.302}$$

Using the stress criterion in the range 2 ($\frac{L}{2} \leq x \leq L$)

$$\sigma_{x,\max} = \frac{M_y(x)}{I_y^{\circledcirc}(x)} \times \frac{d(x)}{2} = \frac{\frac{F_0}{2}(L-x)}{\frac{17}{1250}\pi d^4(x)} \times \frac{d(x)}{2} \stackrel{!}{=} R_{p0.2}, \tag{3.303}$$

results

$$d^3(x) = 2 \times \frac{1250 F_0 (L-x)}{2 \times 4 \times 17 \pi R_{p0.2}} = 2 \times d_0^3 \left(1 - \frac{x}{L}\right). \tag{3.304}$$

In summary, the following distribution results for both sections (see also Fig. 3.18):

$$d(x) = d_0 \left(\frac{2x}{L}\right)^{\frac{1}{3}} \quad \text{for} \quad 0 \leq x \leq \frac{L}{2}, \tag{3.305}$$

$$d(x) = d_0 \left(2\left(1 - \frac{x}{L}\right)\right)^{\frac{1}{3}} \quad \text{for} \quad \frac{L}{2} \leq x \leq L. \tag{3.306}$$

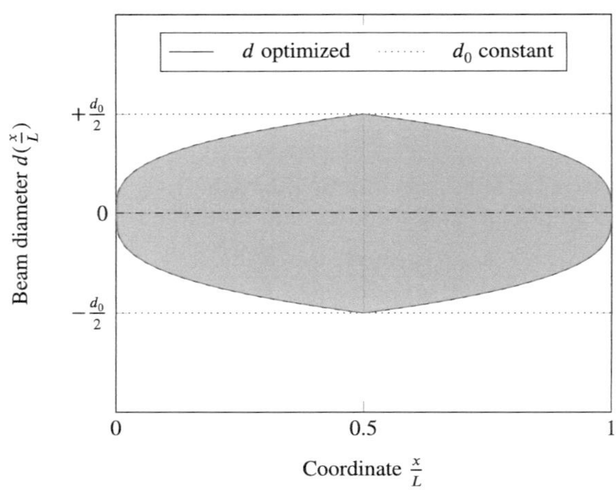

Fig. 3.18 Beam diameter (outer) along the beam axis

- Total strain energy:

The strain energy is calculated according to Eq. (3.32)TB for a bending load case as follows:

$$\Pi = \int_0^L \frac{M_y(x)^2}{2EI_y}\,dx. \tag{3.307}$$

Due to the symmetry of the problem, we get:

$$\Pi = 2 \times \int_0^{\frac{L}{2}} \frac{M_y(x)^2}{2EI_y^{\circledcirc}}\,dx \tag{3.308}$$

$$= \frac{F_0^2}{4 \times \frac{17}{1250} E\pi d_0^4 \left(\frac{2}{L}\right)^{4/3}} \int_0^{\frac{L}{2}} x^{\frac{2}{3}}\,dx \tag{3.309}$$

$$= \frac{375}{272} \times \frac{F_0^2 L^3}{\pi d_0^4 E} \approx 1.379 \times \frac{F_0^2 L^3}{\pi d_0^4 E}. \tag{3.310}$$

- Volume:

$$V = 2 \times \int_0^{\frac{L}{2}} A(x)\,dx \tag{3.311}$$

$$= \frac{8 \times 2^{2/3} \pi d_0^2}{25\,L^{2/3}} \int_0^{\frac{L}{2}} x^{\frac{2}{3}}\,dx \tag{3.312}$$

$$= \frac{12}{125} \times \pi d_0^2 L. \tag{3.313}$$

- Specific energy absorption:

The specific energy absorption is generally given by Eq. (3.27)TB as:

$$SEA = \frac{\Pi}{m}. \tag{3.314}$$

Using the total strain energy according to Eq. (3.310) we get:

$$SEA = \frac{125 \times 375}{12 \times 272} \times \frac{F_0^2 L^2}{\pi^2 d_0^6 \varrho E} \approx 14.361 \frac{F_0^2 L^2}{\pi^2 d_0^6 \varrho E}. \tag{3.315}$$

• Lightweight index:

The definition equation of the lightweight index (see Eq. (3.1)$_{\mathrm{TB}}$) results in:

$$M = \frac{F_0}{F_G}. \tag{3.316}$$

This results in:

$$M = \frac{F_0}{V\varrho g} = \frac{125}{12} \times \frac{F_0}{\pi d_0^2 L\varrho g} \approx 10.417 \times \frac{F_0}{\pi d_0^2 L\varrho g}. \tag{3.317}$$

• Values for beam with constant diameter:

$$\Pi_0 = 2 \times \int_0^{\frac{L}{2}} \frac{M_y(x)^2}{2EI_{y,0}^{\circledcirc}} \mathrm{d}x = \frac{1250}{1632} \times \frac{F_0^2 L^3}{\pi d_0^4 E}, \tag{3.318}$$

$$V_0 = A_0 L = \frac{4}{25} \times \pi d_0^2 L, \tag{3.319}$$

$$SEA_0 = \frac{\Pi_0}{m_0} = \frac{31250}{6528} \times \frac{F_0^2 L^2}{\pi^2 d_0^6 \varrho E} \approx 4.787 \times \frac{F_0^2 L^2}{\pi^2 d_0^6 \varrho E}, \tag{3.320}$$

$$M_0 = \frac{F_0}{F_{G,0}} = \frac{25 F_0}{4\pi d_0^2 L\varrho g}. \tag{3.321}$$

• Comparison:

$$\frac{SEA}{SEA_0} = 3.00, \tag{3.322}$$

$$\frac{M}{M_0} = 1.67. \tag{3.323}$$

2.3.23 **Optimization of a beam under asymmetric 3-point bending**

• Support and internal reactions:

To determine the support reactions, consider the free body diagram shown in Fig. 3.19.

The moment equilibrium around the reference point at the location $x = 0$ allows the determination of the right support force:

$$\sum^{\curvearrowright} M_A = 0: \quad F_0 b - F_B^{\mathrm{R}} L = 0, \tag{3.324}$$

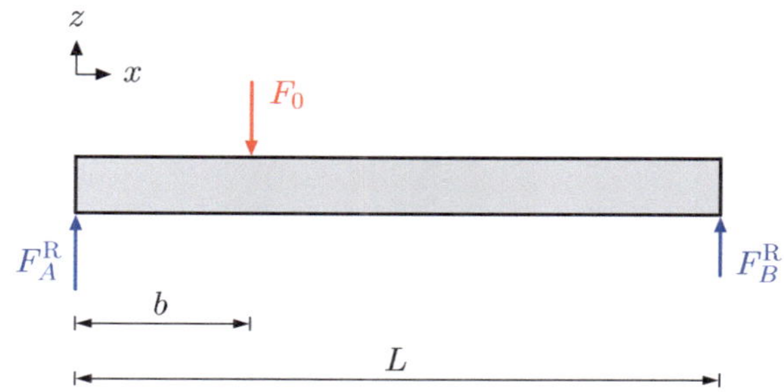

Fig. 3.19 Free body diagram the beam under asymmetric 3-point bending

or solved for the support reaction:

$$F_B^R = F_0 \times \frac{b}{L}. \tag{3.325}$$

The moment equilibrium around the reference point at the location $x = L$ allows the determination of the left support force:

$$\sum^{\curvearrowright} M_B = 0: \quad F_A^R L - F_0(L - b) = 0, \tag{3.326}$$

or solved for the support reaction:

$$F_A^R = F_0 \times \left(1 - \frac{b}{L}\right). \tag{3.327}$$

To determine the internal reactions, the interior must be cut free, see Fig. 3.20.

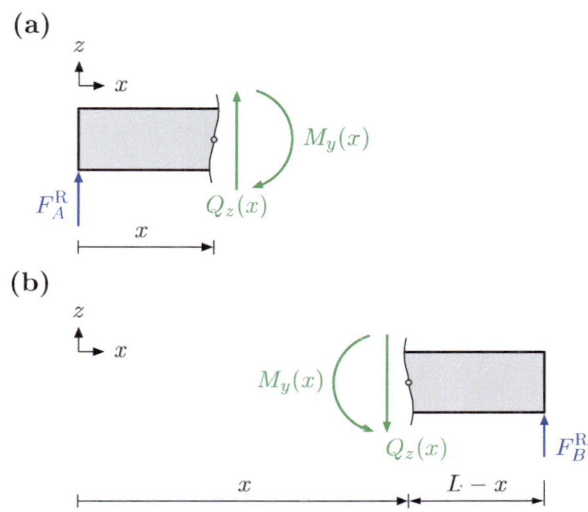

Fig. 3.20 Cutting free inside the beam under asymmetric 3-point bending: **a** Section 1 and **b** Section 2

The moment equilibrium around the reference point at the location x in Section 1 results in (see Fig. 3.20a):

$$\stackrel{\curvearrowright}{\sum} M_y = 0: \quad F_A^R x + M_y(x) = 0, \tag{3.328}$$

respectively the moment equilibrium around the reference point at the location x in Section 2 (see Fig. 3.20b):

$$\stackrel{\curvearrowright}{\sum} M_y = 0: \quad -F_B^R (L - x) - M_y(x) = 0. \tag{3.329}$$

In summary, the internal bending moment distribution, taking into account the expressions for the support reactions according to Eqs. (3.325)–(3.327), can be given as follows:

$$M_y(x) = -F_0 \left(1 - \frac{b}{L}\right) x \text{ for } 0 \leq x \leq b, \tag{3.330}$$

$$M_y(x) = -F_0 \left(1 - \frac{x}{L}\right) b \text{ for } b \leq x \leq L, \tag{3.331}$$

where the maximum moment is at the point of force application:

$$M_{y,\max} = M_y\,(x = b) = -F_0 b \left(1 - \frac{b}{L}\right). \tag{3.332}$$

- Geometric properties of the cross section (I_y, A):

From the solution to the Exercise 2.1.6, i.e.

$$I_y = \frac{1}{12}\left(a_\mathrm{a}^2 + a_\mathrm{i}^2\right)(a_\mathrm{a} - a_\mathrm{i})(a_\mathrm{a} + a_\mathrm{i}), \tag{3.333}$$

$$A = (a_\mathrm{a} - a_\mathrm{i})(a_\mathrm{a} + a_\mathrm{i}), \tag{3.334}$$

follows using $a_\mathrm{o} = d_0$ and $a_\mathrm{i} = d_0 - 2s_0 = \frac{2}{3} d_0$:

$$I_y = \frac{65}{972} d_0^4, \tag{3.335}$$

$$A = \frac{5}{9} d_0^2. \tag{3.336}$$

• Optimized diameter:

Using the stress criterion in the Section 1 ($0 \leq x \leq b$)

$$\sigma_{x,\max} = \frac{M_y(x)}{I_y(x)} \times \frac{d(x)}{2} = \frac{F_0\left(1 - \frac{b}{L}\right)x}{\frac{65}{972}d^4(x)} \times \frac{d(x)}{2} \overset{!}{=} R_{\mathrm{p}0.2}, \tag{3.337}$$

results

$$d^3(x) = \frac{972 F_0 \left(1 - \dfrac{b}{L}\right)}{2 \times 65 R_{\mathrm{p}0.2}} \times x. \tag{3.338}$$

The evaluation of the stress criterion in the reference cross section at the location $x = b$, i.e.

$$\sigma_{x,\max} = \frac{M_{y,\max}}{I_y} \times \frac{d_0}{2} = \frac{F_0\left(1 - \frac{b}{L}\right)b}{\frac{65}{972}d_0^4} \times \frac{d_0}{2} \overset{!}{=} R_{\mathrm{p}0.2}, \tag{3.339}$$

gives the reference diameter as:

$$d_0 = \left(\frac{F_0\left(1 - \frac{b}{L}\right)b}{2 \times \frac{65}{972} R_{\mathrm{p}0.2}}\right)^{\frac{1}{3}}. \tag{3.340}$$

With this reference diameter, the following results for Section 1:

$$d^3(x) = d_0^3 \times \frac{x}{b}. \tag{3.341}$$

Using the stress criterion in the Section 2 ($b \leq x \leq L$)

$$\sigma_{x,\max} = \frac{M_y(x)}{I_y(x)} \times \frac{d(x)}{2} = \frac{F_0 b\left(1 - \frac{x}{L}\right)}{\frac{65}{972}d^4(x)} \times \frac{d(x)}{2} \overset{!}{=} R_{\mathrm{p}0.2}, \tag{3.342}$$

results

$$d^3(x) = \frac{972 F_0 b\left(1 - \frac{b}{L}\right)}{2 \times 65 R_{\mathrm{p}0.2}} \times \frac{1 - \dfrac{x}{L}}{1 - \dfrac{b}{L}} = d_0^3 \times \frac{1 - \dfrac{x}{L}}{1 - \dfrac{b}{L}}. \tag{3.343}$$

In summary, the following distribution results for both sections (see also Fig. 3.21):

$$d(x) = d_0 \left(\frac{x}{b}\right)^{\frac{1}{3}} \quad \text{for } x \leq b, \tag{3.344}$$

$$d(x) = d_0 \left(\frac{1 - \frac{x}{L}}{1 - \frac{b}{L}}\right)^{\frac{1}{3}} \quad \text{for } x \geq b. \tag{3.345}$$

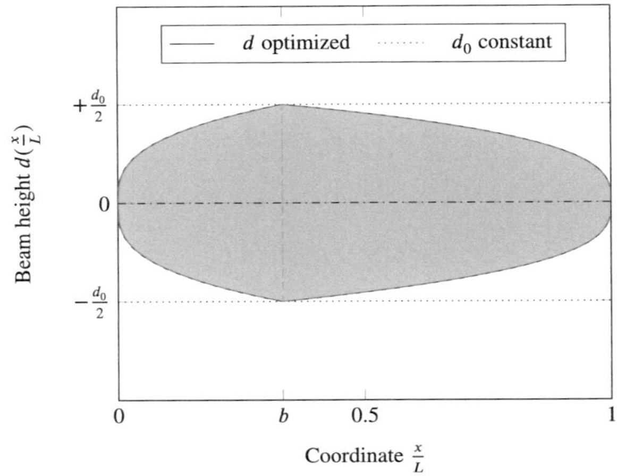

Fig. 3.21 Beam height along the beam axis

• Reference cross section:

The expressions for A_0, I_0, and F_0 can be found in Eqs. (3.428), (3.427), and (3.340). The defining equation of the lightweight index (see Eq. $(3.1)_{\text{TB}}$) is:

$$M = \frac{F_0}{F_G}, \tag{3.346}$$

or with the corresponding quantities for the reference cross section:

$$M_0 = \frac{\frac{65 d_0^3 R_{\text{p0.2}}}{486 b(1-b/L)}}{\frac{5}{9}\varrho g d_0^2 L}. \tag{3.347}$$

The total strain energy is calculated according to Eq. $(3.32)_{\text{TB}}$ for a bending load case as follows:

$$\Pi = \int_0^L \frac{M_y(x)^2}{2EI_y} \mathrm{d}x. \tag{3.348}$$

Due to the area subdivision, this results in:

$$\Pi_0 = \int_0^b \frac{M_y(x)^2}{2EI_y} \mathrm{d}x + \int_b^L \frac{M_y(x)^2}{2EI_y} \mathrm{d}x, \tag{3.349}$$

$$= \frac{1}{2E} \int_0^b \frac{F_0^2 \left(1 - \frac{b}{L}\right)^2 x^2}{\frac{65}{972} d_0^4} \mathrm{d}x + \frac{1}{2E} \int_b^L \frac{F_0^2 \left(\frac{b}{L}\right)^2 (L-x)^2}{\frac{65}{972} d_0^4} \mathrm{d}x, \tag{3.350}$$

$$= \frac{1}{6E} \times \frac{F_0^2 b^3}{\frac{65}{972} d_0^4} \times \left(1 - \frac{b}{L}\right)^2 \times \left(\frac{L}{b}\right). \tag{3.351}$$

In summary, one obtains for the reference cross section:

$$A_0 = \frac{5}{9} d_0^2, \tag{3.352}$$

$$I_0 = \frac{65}{972} d_0^4, \tag{3.353}$$

$$F_0 = \frac{65 d_0^3 R_{\text{p0.2}}}{486 b(1 - b/L)}, \tag{3.354}$$

$$M_0 = \frac{13 d_0 R_{\text{p0.2}}}{54 \varrho g L b(1 - b/L)}, \tag{3.355}$$

$$\Pi_0 = \frac{65 d_0^2 L R_{\text{p0.2}}^2}{1458 E}, \tag{3.356}$$

$$SEA_0 = \frac{13 R_{\text{p0.2}}^2}{162 \varrho E}. \tag{3.357}$$

- Optimized cross section:

The expressions for $A(x)$, $I(x)$, and F_0 can be adapted from the expressions of the reference cross section. To calculate the volume, the area subdivision must be taken into account:

$$V = \int_0^b A(x)\, \mathrm{d}x + \int_b^L A(x)\, \mathrm{d}x, \tag{3.358}$$

$$= \frac{5}{9} \times \frac{d_0^2}{b^{\frac{2}{3}}} \int_0^b x^{\frac{2}{3}}\, \mathrm{d}x + \frac{5}{9} \times \frac{d_0^2}{\left(1 - \frac{b}{L}\right)^{\frac{2}{3}}} \int_b^L \left(1 - \frac{x}{L}\right)^{\frac{2}{3}} \mathrm{d}x \tag{3.359}$$

$$= \frac{d_0^2 L}{3}. \tag{3.360}$$

The calculation of the specific strain energy must be carried out accordingly:

$$\Pi = \int_0^b \frac{M_y(x)^2}{2EI_y(x)} \mathrm{d}x + \int_b^L \frac{M_y(x)^2}{2EI_y(x)} \mathrm{d}x, \tag{3.361}$$

$$= \frac{1}{2E} \int_0^b \frac{F_0^2 \left(1 - \frac{b}{L}\right)^2 x^2}{\frac{65}{972} d_0^4 \left(\frac{x}{b}\right)^{\frac{4}{3}}} \mathrm{d}x + \frac{1}{2E} \int_b^L \frac{F_0^2 \left(\frac{b}{L}\right)^2 (L - x)^2}{\frac{65}{972} d_0^4 \left(\frac{1 - x/L}{1 - b/L}\right)^{\frac{4}{3}}} \mathrm{d}x, \tag{3.362}$$

$$= \frac{1458}{325} \times \frac{F_0^2 b^2 L \left(1 - \frac{b}{L}\right)^2}{d_0^4 E}. \tag{3.363}$$

In summary, one obtains for the optimized cross section:

$$A(x) = \frac{5}{9}d(x)^2, \tag{3.364}$$

$$I(x) = \frac{65}{972}d(x)^4, \tag{3.365}$$

$$V = \frac{1}{3}d_0^2 L, \tag{3.366}$$

$$F_0 = \frac{65 d_0^3 R_{\text{p0.2}}}{486 b(1 - b/L)}, \tag{3.367}$$

$$M = \frac{65 d_0 R_{\text{p0.2}}}{162 \varrho g L b(1 - b/L)}, \tag{3.368}$$

$$\Pi = \frac{13 d_0^2 L R_{\text{p0.2}}^2}{162 E}, \tag{3.369}$$

$$SEA = \frac{13 R_{\text{p0.2}}^2}{54 \varrho E}. \tag{3.370}$$

- Comparison:

$$\frac{m}{m_0} = \frac{3}{5}, \tag{3.371}$$

$$\frac{M}{M_0} = \frac{5}{3}, \tag{3.372}$$

$$\frac{SEA}{SEA_0} = 3. \tag{3.373}$$

Numerical values for specific b-values are shown in Table 3.3.

Table 3.3 Comparison of the lightweight index and the specific energy absorption for different values of b

$\frac{b}{L}$	$\frac{M_0}{\frac{d_0 R_{\text{p0.2}}}{\varrho g L^2}}$	$\frac{M}{\frac{d_0 R_{\text{p0.2}}}{\varrho g L^2}}$	$\frac{SEA_0}{\frac{R_{\text{p0.2}}^2}{E\varrho}}$	$\frac{SEA}{\frac{R_{\text{p0.2}}^2}{E\varrho}}$	$\frac{M_0}{M}$	$\frac{SEA_0}{SEA}$
$\frac{1}{4}$	$\frac{104}{81}$	$\frac{520}{243}$	$\frac{13}{162}$	$\frac{13}{54}$	$\frac{5}{3}$	3
$\frac{1}{3}$	$\frac{13}{12}$	$\frac{65}{36}$	$\frac{13}{162}$	$\frac{13}{54}$	$\frac{5}{3}$	3
$\frac{1}{2}$	$\frac{26}{27}$	$\frac{130}{81}$	$\frac{13}{162}$	$\frac{13}{54}$	$\frac{5}{3}$	3

2.3.24 **Optimization of a simply supported beam under constant distributed load: rectangular cross section**

• Moment distribution:

The bending moment distribution can be taken from the solution of Exercise 2.2.12:

$$M_y(x) = \frac{q_0}{2}(x^2 - Lx) = \frac{q_0 L^2}{2}\left(\left[\frac{x}{L}\right]^2 - \left[\frac{x}{L}\right]^1\right) < 0. \tag{3.374}$$

The maximum bending moment is obtained in the middle of the beam as:

$$M_{y,\max} = M_y\left(\tfrac{L}{2}\right) = -\frac{q_0 L^2}{8}. \tag{3.375}$$

• Geometric properties of the cross section:

For a rectangular cross section of width b and height h we get:

$$I_y = \frac{bh^3}{12}, \tag{3.376}$$

$$A = bh. \tag{3.377}$$

• Optimized diameter:

Using the stress criterion

$$\sigma_{x,\max} = \frac{M_y(x)}{I_y(x)} \times \frac{h(x)}{2} = \frac{\frac{q_0 L^2}{2}\left(\left[\frac{x}{L}\right] - \left[\frac{x}{L}\right]^2\right)}{\frac{bh(x)^3}{12}} \times \frac{h(x)}{2} \stackrel{!}{=} R_{\mathrm{p}0.2}, \tag{3.378}$$

results

$$h^2(x) = \frac{q_0 L^2\left(\left[\frac{x}{L}\right] - \left[\frac{x}{L}\right]^2\right)}{4 \times \frac{1}{12} b R_{\mathrm{p}0.2}}. \tag{3.379}$$

The evaluation of the stress criterion in the reference cross section at the location $x = L/2$, i.e.

$$\sigma_{x,\max} = \frac{M_{y,\max}}{I_y} \times \frac{h_0}{2} = \frac{\frac{q_0 L^2}{8}}{\frac{bh_0^3}{12}} \times \frac{h_0}{2} \stackrel{!}{=} R_{\mathrm{p}0.2}, \tag{3.380}$$

returns the reference height to:

$$h_0^2 = \frac{q_0 L^2}{16 \times \frac{b}{12} R_{\mathrm{p}0.2}}. \tag{3.381}$$

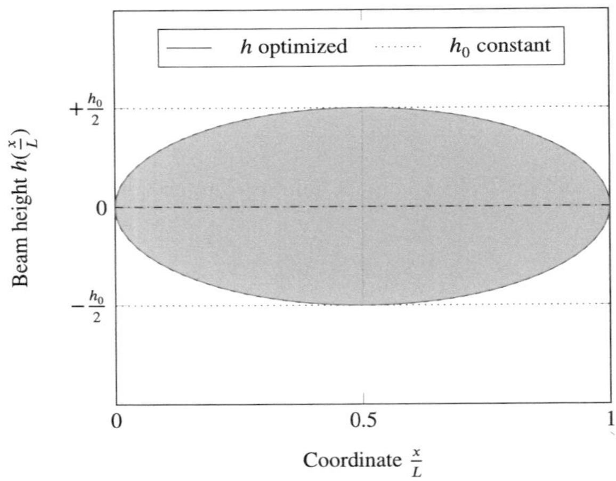

Fig. 3.22 Beam height along the beam axis

With this result, the optimized cross-sectional height is finally calculated as (see also Fig. 3.22):

$$h(x) = 2h_0 \left(\left[\frac{x}{L}\right] - \left[\frac{x}{L}\right]^2 \right)^{\frac{1}{2}}. \tag{3.382}$$

• Optimized diameter:
The volume is determined by evaluating the integral:

$$V = \int_0^L A(x)\,\mathrm{d}x = 2bh_0 \underbrace{\int_0^L \left(\left[\frac{x}{L}\right] - \left[\frac{x}{L}\right]^2 \right)^{\frac{1}{2}} \mathrm{d}x}_{\frac{\pi L}{8}} = \frac{1}{4} \times \pi b h_0 L. \tag{3.383}$$

The strain energy is calculated according to Eq. (3.32)$_{\mathrm{TB}}$ for a bending load case as follows:

$$\Pi = \int_0^L \frac{M_y(x)^2}{2EI_y}\mathrm{d}x. \tag{3.384}$$

For our example we get:

$$\Pi = \frac{1}{2E}\int_0^L \frac{\frac{q_0^2 L^4}{4}\left(\left[\frac{x}{L}\right]-\left[\frac{x}{L}\right]^2\right)^2}{\frac{1}{12}b2^3h_0^3\left(\left[\frac{x}{L}\right]-\left[\frac{x}{L}\right]^2\right)^{\frac{3}{2}}}\mathrm{d}x \tag{3.385}$$

$$= \frac{3q_0^2L^4}{16Ebh_0^3}\underbrace{\int_0^L\left(\left[\frac{x}{L}\right]-\left[\frac{x}{L}\right]^2\right)^{\frac{1}{2}}\mathrm{d}x}_{\frac{\pi L}{8}} \tag{3.386}$$

$$= \frac{3\pi}{128}\times\frac{L^5q_0^2}{Ebh_0^3}. \tag{3.387}$$

From the relationship for the stress criterion at the location $x = L/2$ (see Eq. (3.381)) the expression for the distributed load is:

$$q_0 = \frac{16bh_0^2R_{\mathrm{p0.2}}}{12\,L^2}. \tag{3.388}$$

In summary, one obtains for the optimized cross section:

$$V = \frac{b\pi Lh_0}{4}, \tag{3.389}$$

$$\Pi = \frac{3\pi L^5q_0^2}{128Ebh_0^3}, \tag{3.390}$$

$$SEA = \frac{3}{32}\times\frac{L^4q_0^2}{Eb^2h_0^4\varrho} = \frac{1}{6}\times\frac{R_{\mathrm{p0.2}}^2}{E\varrho}. \tag{3.391}$$

• Reference cross section:

The total strain energy for the reference cross section is obtained by integration as:

$$\Pi_0 = \frac{1}{2E}\int_0^L \frac{\frac{q_0^2L^4}{4}\left(\left[\frac{x}{L}\right]-\left[\frac{x}{L}\right]^2\right)^2}{\frac{1}{12}bh_0^3}\mathrm{d}x, \tag{3.392}$$

$$= \frac{3q_0^2L^4}{2Ebh_0^3}\underbrace{\int_0^L\left(\left[\frac{x}{L}\right]^2-2\left[\frac{x}{L}\right]^3+\left[\frac{x}{L}\right]^4\right)\mathrm{d}x}_{\frac{L}{30}}, \tag{3.393}$$

$$= \frac{q_0^2L^5}{20Ebh_0^3}. \tag{3.394}$$

In summary, one obtains for the reference cross section:

$$V_0 = bh_0L, \tag{3.395}$$

$$\Pi_0 = \frac{L^5 q_0^2}{20Ebh_0^3}, \tag{3.396}$$

$$SEA_0 = \frac{1}{20} \times \frac{L^4 q_0^2}{Eb^2 h_0^4 \varrho} = \frac{4}{45} \times \frac{R_{\text{p0.2}}^2}{E\varrho}. \tag{3.397}$$

• Comparison:

$$\frac{SEA}{SEA_0} = \frac{15}{8} \approx 1.875. \tag{3.398}$$

2.3.25 **Optimization of a simply supported beam under constant distributed load: tubular cross section (numerical integration)**

• Moment distribution:

The bending moment distribution can be taken from the solution of Exercise 2.2.12:

$$M_y(x) = \frac{q_0}{2}(x^2 - Lx) = \frac{q_0 L^2}{2}\left(\left[\frac{x}{L}\right]^2 - \left[\frac{x}{L}\right]^1\right) < 0. \tag{3.399}$$

The maximum bending moment is obtained in the middle of the beam as:

$$M_{y,\max} = M_y\left(\tfrac{L}{2}\right) = -\frac{q_0 L^2}{8}. \tag{3.400}$$

• Geometric properties of the cross section (I_y, A):

From the solution to the Exercise 2.1.5, i.e.

$$I_y = \frac{\pi}{4}\left(r_\text{o}^2 + r_\text{i}^2\right)(r_\text{o} - r_\text{i})(r_\text{o} + r_\text{i}), \tag{3.401}$$

$$A = \pi\,(r_\text{o} - r_\text{i})(r_\text{o} + r_\text{i}), \tag{3.402}$$

follows using $r_\text{o} = d_0/2$ and $r_\text{i} = \frac{d_0}{2} - s_0 = \frac{1}{3}\,d_0$:

$$I_y = \frac{65\pi}{5184}d_0^4, \tag{3.403}$$

$$A = \frac{5\pi}{36}d_0^2. \tag{3.404}$$

• Optimized diameter:

Using the stress criterion

$$\sigma_{x,\max} = \frac{M_y(x)}{I_y(x)} \times \frac{d(x)}{2} = \frac{\frac{q_0 L^2}{2}\left(\left[\frac{x}{L}\right] - \left[\frac{x}{L}\right]^2\right)}{\frac{65\pi d(x)^4}{5184}} \times \frac{h(x)}{2} \stackrel{!}{=} R_{\text{p0.2}}, \tag{3.405}$$

results

$$d^3(x) = \frac{1296 q_0 L^2}{65\pi R_{\text{p0.2}}} \times \left(\left[\frac{x}{L}\right] - \left[\frac{x}{L}\right]^2\right). \tag{3.406}$$

The evaluation of the stress criterion in the reference cross-section at the location $x = L/2$, i.e.

$$\sigma_{x,\max} = \frac{M_{y,\max}}{I_y} \times \frac{d_0}{2} = \frac{\frac{q_0 L^2}{8}}{\frac{65\pi d_0^4}{5184}} \times \frac{d_0}{2} \stackrel{!}{=} R_{\text{p0.2}}, \tag{3.407}$$

results in the reference diameter as:

$$d_0^3 = \frac{324 q_0 L^2}{65\pi R_{\text{p0.2}}}. \tag{3.408}$$

With this result, the optimized diameter is finally calculated as (see also Fig. 3.23):

$$d(x) = 4^{\frac{1}{3}} d_0 \left(-\left[\frac{x}{L}\right]^2 + \left[\frac{x}{L}\right]^1\right)^{\frac{1}{3}}, \tag{3.409}$$

with

$$d_0 = \left(\frac{324 q_0 L^2}{65\pi R_{\text{p0.2}}}\right)^{\frac{1}{3}} = 35.17\,\text{mm}. \tag{3.410}$$

The volume is determined by evaluating the integral:

$$V = \int_0^L A(x)\,\mathrm{d}x = \frac{5 \times 4^{\frac{2}{3}} \pi d_0^2}{36} \underbrace{\int_0^L \left(\left[\frac{x}{L}\right] - \left[\frac{x}{L}\right]^2\right)^{\frac{2}{3}} \mathrm{d}x}_{0.2933414631064508\,L}. \tag{3.411}$$

The strain energy is calculated according to Eq. $(3.32)_{\text{TB}}$ for a bending load case as follows:

$$\Pi = \int_0^L \frac{M_y(x)^2}{2EI_y}\,\mathrm{d}x. \tag{3.412}$$

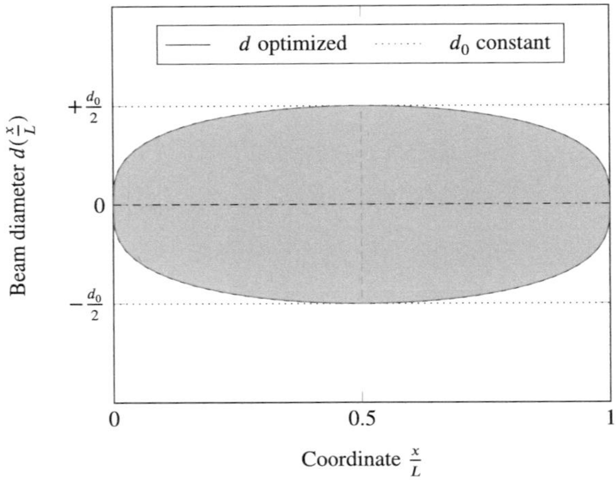

Fig. 3.23 Beam diameter along the beam axis

For our example we get:

$$\Pi = \frac{1}{2E}\int_0^L \frac{\frac{q_0^2L^4}{4}\left(\left[\frac{x}{L}\right]-\left[\frac{x}{L}\right]^2\right)^2}{\frac{65\pi}{5184}4^{4/3}d_0^4\left(\left[\frac{x}{L}\right]-\left[\frac{x}{L}\right]^2\right)^{\frac{4}{3}}}\mathrm{d}x \tag{3.413}$$

$$= \frac{5184q_0^2L^4}{8\times 65\times 4^{4/3}E\pi d_0^4}\underbrace{\int_0^L\left(\left[\frac{x}{L}\right]-\left[\frac{x}{L}\right]^2\right)^{\frac{2}{3}}\mathrm{d}x}_{0.2933414631064508\,L}. \tag{3.414}$$

The numerical values finally result in:

$$V = 1013504.27\,\mathrm{mm}^3, \tag{3.415}$$

$$\Pi = 161921.87\,\mathrm{Nmm}. \tag{3.416}$$

- Constant diameter:

The total strain energy for the reference cross section is obtained by integration as:

$$\Pi_0 = \frac{1}{2E}\int_0^L \frac{\frac{q_0^2L^4}{4}\left(\left[\frac{x}{L}\right]-\left[\frac{x}{L}\right]^2\right)^2}{\frac{65\pi}{5184}d_0^4}\mathrm{d}x, \tag{3.417}$$

$$= \frac{5184q_0^2L^4}{8\times 65E\pi d_0^4}\underbrace{\int_0^L\left(\left[\frac{x}{L}\right]^2-2\left[\frac{x}{L}\right]^3+\left[\frac{x}{L}\right]^4\right)\mathrm{d}x}_{\frac{L}{30}}, \tag{3.418}$$

$$= \frac{5184 q_0^2 L^5}{15600 E \pi d_0^4}. \tag{3.419}$$

In summary, one obtains for the reference beam:

$$V_0 = 1371130.53\,\mathrm{mm}^3, \tag{3.420}$$

$$\Pi_0 = 116831.68\,\mathrm{Nmm}. \tag{3.421}$$

• Ratio:

$$\frac{SEA}{SEA_0} = 1.875. \tag{3.422}$$

2.3.26 **Optimization of a beam with a square box profile along the longitudinal axis under the influence of a constant distributed load**

• Moment distribution:

The bending moment distribution can be taken from the solution of Exercise 2.2.12:

$$M_y(x) = \frac{q_0}{2}(x^2 - Lx) = \frac{q_0 L^2}{2}\left(\left[\frac{x}{L}\right]^2 - \left[\frac{x}{L}\right]^1\right) < 0. \tag{3.423}$$

The maximum bending moment is obtained in the middle of the beam as:

$$M_{y,\max} = M_y\left(\tfrac{L}{2}\right) = -\frac{q_0 L^2}{8}. \tag{3.424}$$

• Geometric properties of the cross section (I_y, A):

From the solution to the Exercise 2.1.6, i.e.

$$I_y = \frac{1}{12}\left(a_o^2 + a_i^2\right)(a_o - a_i)(a_o + a_i), \tag{3.425}$$

$$A = (a_o - a_i)(a_o + a_i), \tag{3.426}$$

follows using $a_o = d_0$ and $a_i = d_0 - 2s_0 = \frac{2}{3}d_0$:

$$I_y = \frac{65}{972} d_0^4, \tag{3.427}$$

$$A = \frac{5}{9} d_0^2. \tag{3.428}$$

- Reference beam with constant cross section:

The definition equation of the lightweight index (see Eq. $(3.1)_{\text{TB}}$) results in:

$$M = \frac{F_0}{F_G}. \tag{3.429}$$

This results with the approximation $F_0 \approx q_0 L$:

$$M_0 \approx \frac{q_0 L}{\varrho g A_0 L}. \tag{3.430}$$

The evaluation of the stress criterion in the reference cross section at the location $x = L/2$, i.e.

$$\sigma_{x,\max} = \frac{M_{y,\max}}{I_y} \times \frac{d_0}{2} = \frac{\frac{q_0 L^2}{8}}{\frac{65 d_0^4}{972}} \times \frac{d_0}{2} \overset{!}{=} R_{\text{p}0.2}, \tag{3.431}$$

gives the expression for the distributed load as:

$$q_0 = \frac{260}{243} \times \frac{d_0^3 R_{\text{p}0.2}}{L^2}. \tag{3.432}$$

This gives the lightweight index as:

$$M_0 \approx \frac{52}{27} \times \frac{d_0 R_{\text{p}0.2}}{\varrho g L^2}. \tag{3.433}$$

The total strain energy for the reference cross section is obtained by integration to:

$$\Pi_0 = \frac{1}{2E} \int_0^L \frac{\frac{q_0^2 L^4}{4} \left(\left[\frac{x}{L}\right] - \left[\frac{x}{L}\right]^2 \right)^2}{\frac{65}{972} d_0^4} \mathrm{d}x, \tag{3.434}$$

$$= \frac{972 q_0^2 L^4}{8 \times 65 E d_0^4} \underbrace{\int_0^L \left(\left[\frac{x}{L}\right]^2 - 2\left[\frac{x}{L}\right]^3 + \left[\frac{x}{L}\right]^4 \right) \mathrm{d}x}_{\frac{L}{30}}, \tag{3.435}$$

$$= \frac{972 q_0^2 L^5}{15600 E d_0^4}. \tag{3.436}$$

In summary, one obtains for the reference beam:

$$A_0 = \frac{5d_0^2}{9}, \tag{3.437}$$

$$I_0 = \frac{65d_0^4}{972}, \tag{3.438}$$

$$q_0 = \frac{260d_0^3 R_{p0.2}}{243L^2}, \tag{3.439}$$

$$M_0 \approx \frac{52}{27} \times \frac{d_0 R_{p0.2}}{\varrho g L^2}, \tag{3.440}$$

$$\approx 1.92593 \times \frac{d_0 R_{p0.2}}{\varrho g L^2}, \tag{3.441}$$

$$\Pi_0 = \frac{52}{729} \times \frac{L d_0^2 R_{p0.2}^2}{E}, \tag{3.442}$$

$$SEA_0 = \frac{52}{405} \times \frac{R_{p0.2}^2}{\varrho E}, \tag{3.443}$$

$$\approx 0.12840 \times \frac{R_{p0.2}^2}{\varrho E}. \tag{3.444}$$

- Beam with optimized cross section:

Using the stress criterion

$$\sigma_{x,\max} = \frac{M_y(x)}{I_y(x)} \times \frac{d(x)}{2} = \frac{\frac{q_0 L^2}{2}\left(\left[\frac{x}{L}\right] - \left[\frac{x}{L}\right]^2\right)}{\frac{65d(x)^4}{972}} \times \frac{h(x)}{2} \overset{!}{=} R_{p0.2}. \tag{3.445}$$

results

$$d^3(x) = \frac{972 q_0 L^2}{1260 R_{p0.2}} \times \left(\left[\frac{x}{L}\right] - \left[\frac{x}{L}\right]^2\right). \tag{3.446}$$

The evaluation of the stress criterion in the reference cross section at the location $x = L/2$ provides the reference dimension (see Eq. (3.432)):

$$d_0^3 = \frac{243 q_0 L^2}{260 R_{p0.2}}. \tag{3.447}$$

With this result we get:

$$d^3(x) = 4 \times \underbrace{\frac{243 q_0 L^2}{260 R_{p0.2}}}_{d_0^3} \times \left(\left[\frac{x}{L}\right] - \left[\frac{x}{L}\right]^2\right). \tag{3.448}$$

The volume is determined by evaluating the integral:

$$V = \int_0^L A(x)\,\mathrm{d}x = \frac{5 \times 4^{\frac{2}{3}} d_0^2}{9} \underbrace{\int_0^L \left(\left[\frac{x}{L}\right] - \left[\frac{x}{L}\right]^2 \right)^{\frac{2}{3}} \mathrm{d}x}_{0.2933414631064508\,L}. \tag{3.449}$$

The lightweight index is again approximately:

$$M \approx \frac{q_0 L}{\varrho g V} = \frac{q_0 L}{\varrho g \times \frac{5}{9} \times 4^{2/3} d_0^2 \times 0.293341\,L}. \tag{3.450}$$

The total strain energy is:

$$\Pi = \frac{1}{2E} \int_0^L \frac{\frac{q_0^2 L^4}{4} \left(\left[\frac{x}{L}\right] - \left[\frac{x}{L}\right]^2 \right)^2}{\frac{65}{972} 4^{4/3} d_0^4 \left(\left[\frac{x}{L}\right] - \left[\frac{x}{L}\right]^2 \right)^{\frac{4}{3}}} \mathrm{d}x \tag{3.451}$$

$$= \frac{972 q_0^2 L^4}{8 \times 65 \times 4^{4/3} E d_0^4} \underbrace{\int_0^L \left(\left[\frac{x}{L}\right] - \left[\frac{x}{L}\right]^2 \right)^{\frac{2}{3}} \mathrm{d}x}_{0.2933414631064508\,L}. \tag{3.452}$$

In summary, one obtains for the optimized cross section:

$$d(x) = 4^{\frac{1}{3}} d_0 \left(\left[\frac{x}{L}\right]^1 - \left[\frac{x}{L}\right]^2 \right)^{\frac{1}{3}}, \tag{3.453}$$

$$d_0 = \frac{1}{4^{\frac{1}{3}}} \left(\frac{243}{65} \times \frac{q_0 L^2}{R_{\mathrm{p}0.2}} \right)^{\frac{1}{3}}, \tag{3.454}$$

$$V = \frac{5}{9} \times 4^{\frac{2}{3}} \times 0.293342 d_0^2 L, \tag{3.455}$$

$$\approx 0.41 \times d_0^2 L, \tag{3.456}$$

$$M \approx \frac{52}{27} \times \frac{1}{4^{\frac{2}{3}} \times 0.293342} \times \frac{d_0 R_{\mathrm{p}0.2}}{\varrho g L^2}, \tag{3.457}$$

$$\approx 2.60551 \times \frac{d_0 R_{\mathrm{p}0.2}}{\varrho g L^2}, \tag{3.458}$$

$$\Pi = \frac{2 \times 52 \times 260 \times 0.293342}{4^{\frac{7}{3}} \times 13 \times 243} \times \frac{L d_0^2 R_{\mathrm{p}0.2}^2}{E}, \tag{3.459}$$

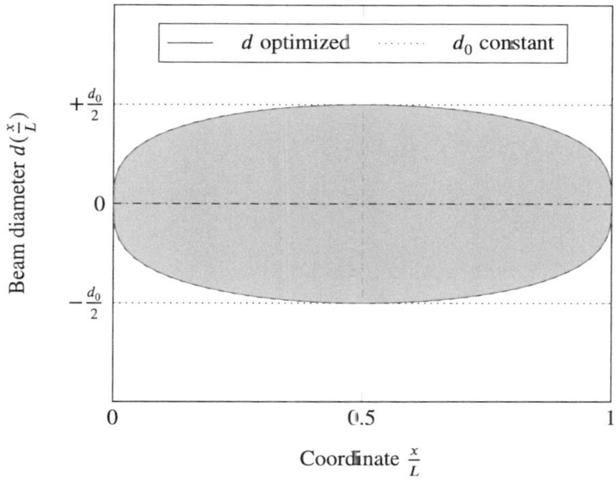

Fig. 3.24 Beam diameter along the beam axis

$$SEA = \frac{13}{54} \times \frac{R_{p0.2}^2}{\varrho E} \approx 0.24074 \times \frac{R_{p0.2}^2}{\varrho E}. \tag{3.460}$$

The optimized beam is shown in Fig. 3.24.

2.3.27 **Optimization of a beam with a rectangular box profile along the longitudinal axis under the influence of a constant distributed load**

• Moment distribution:

The bending moment distribution can be taken from the solution of Exercise 2.2.12:

$$M_y(x) = \frac{q_0}{2}(x^2 - Lx) = \frac{q_0 L^2}{2}\left(\left[\frac{x}{L}\right]^2 - \left[\frac{x}{L}\right]^1\right) < 0. \tag{3.461}$$

The maximum bending moment is obtained in the middle of the beam as:

$$M_{y,\max} = M_y\left(\tfrac{L}{2}\right) = -\frac{q_0 L^2}{8}. \tag{3.462}$$

• Geometric parameters of the cross section (I_y, A):

For a box profile with external dimensions $a \times 2a$ and constant wall thickness $d = a/6$ the following results:

$$I_y = \frac{1}{12}a(2a)^3 - \frac{1}{12}(a - 2d)(2a - 2d)^3 = \frac{199}{486}a^4, \tag{3.463}$$

$$A = 2a^2 - (a - 2d)(2a - 2d) = \frac{8}{9}a^2. \tag{3.464}$$

• Reference beam with constant cross section:

The evaluation of the stress criterion in the reference cross section at the location $x = L/2$, i.e.

$$\sigma_{x,\max} = \frac{M_{y,\max}}{I_y} \times a_0 = \frac{\frac{q_0 L^2}{8}}{\frac{199 a_0^4}{486}} \times a_0 \overset{!}{=} R_{\text{p0.2}}, \tag{3.465}$$

results the expression for the distributed load as:

$$q_0 = \frac{796}{243} \times \frac{a_0^3 R_{\text{p0.2}}}{L^2}. \tag{3.466}$$

The definition equation of the lightweight index (see Eq. $(3.1)_{\text{TB}}$) results in:

$$M = \frac{F_0}{F_{\text{G}}}. \tag{3.467}$$

This results with the approximation $F_0 \approx q_0 L$:

$$M_0 \approx \frac{q_0 L}{\varrho g A_0 L} = \frac{q_0 L}{\varrho g L \frac{8}{9} a_0^2} = \frac{199}{54} \times \frac{a_0 R_{\text{p0.2}}}{\varrho g L^2}. \tag{3.468}$$

The total strain energy for the reference cross section is obtained by integration as:

$$\Pi_0 = \frac{1}{2E} \int_0^L \frac{\frac{q_0^2 L^4}{4} \left(\left[\frac{x}{L}\right] - \left[\frac{x}{L}\right]^2 \right)^2}{\frac{199}{486} a_0^4} \mathrm{d}x, \tag{3.469}$$

$$= \frac{486 q_0^2 L^4}{2 \times 199 \times 4 E a_0^4} \underbrace{\int_0^L \left(\left[\frac{x}{L}\right]^2 - 2 \left[\frac{x}{L}\right]^3 + \left[\frac{x}{L}\right]^4 \right) \mathrm{d}x}_{\frac{L}{30}}, \tag{3.470}$$

$$= \frac{243 q_0^2 L^5}{199 \times 4 \times 30 E a_0^4} = \frac{398}{3645} \times \frac{L a_0^2 R_{\text{p0.2}}^2}{E}. \tag{3.471}$$

In summary, one obtains for the reference beam:

$$A_0 = \frac{8 a_0^2}{9}, \tag{3.472}$$

$$I_0 = \frac{199 a_0^4}{486}, \tag{3.473}$$

$$q_0 = \frac{796 a_0^3 R_{\text{p0.2}}}{243 L^2}, \tag{3.474}$$

$$M_0 \approx \frac{199}{54} \times \frac{a_0 R_{p0.2}}{\varrho g L^2}, \tag{3.475}$$

$$\approx 3.685185 \times \frac{a_0 R_{p0.2}}{\varrho g L^2}, \tag{3.476}$$

$$\Pi_0 = \frac{398}{3645} \times \frac{L a_0^2 R_{p0.2}^2}{E}, \tag{3.477}$$

$$\approx 0.109191 \times \frac{L a_0^2 R_{p0.2}^2}{E}, \tag{3.478}$$

$$SEA_0 = \frac{199}{1620} \times \frac{R_{p0.2}^2}{\varrho E}, \tag{3.479}$$

$$\approx 0.122840 \times \frac{R_{p0.2}^2}{\varrho E}. \tag{3.480}$$

- Beam with optimized cross section:

Using the stress criterion

$$\sigma_{x,\max} = \frac{M_y(x)}{I_y(x)} \times a(x) = \frac{\frac{q_0 L^2}{2}\left(\left[\frac{x}{L}\right] - \left[\frac{x}{L}\right]^2\right)}{\frac{199 a(x)^4}{486}} \times a(x) \stackrel{!}{=} R_{p0.2}, \tag{3.481}$$

results

$$a^3(x) = \frac{243 q_0 L^2}{199 R_{p0.2}} \times \left(\left[\frac{x}{L}\right] - \left[\frac{x}{L}\right]^2\right). \tag{3.482}$$

The evaluation of the stress criterion in the reference cross section at the location $x = L/2$ yields the reference dimension (see Eq. (3.466)):

$$a_0^3 = \frac{243 q_0 L^2}{796 R_{p0.2}}. \tag{3.483}$$

With this result we get:

$$a^3(x) = 4 \times \underbrace{\frac{243 q_0 L^2}{796 R_{p0.2}}}_{a_0^3} \times \left(\left[\frac{x}{L}\right] - \left[\frac{x}{L}\right]^2\right). \tag{3.484}$$

The volume is determined by evaluating the integral:

$$V = \int_0^L A(x)\,dx = \frac{8 \times 4^{\frac{2}{3}} a_0^2}{9} \underbrace{\int_0^L \left(\left[\frac{x}{L}\right] - \left[\frac{x}{L}\right]^2\right)^{\frac{2}{3}} dx}_{0.2933414631064508\,L}. \tag{3.485}$$

The lightweight index is again approximately:

$$M \approx \frac{q_0 L}{\varrho g V} = \frac{q_0 L}{\varrho g \times \frac{8}{9} \times 4^{2/3} a_0^2 \times 0.293341\, L}. \tag{3.486}$$

The total strain energy is:

$$\Pi = \frac{1}{2E} \int_0^L \frac{\frac{q_0^2 L^4}{4} \left(\left[\frac{x}{L}\right] - \left[\frac{x}{L}\right]^2 \right)^2}{\frac{199}{486} 4^{4/3} a_0^4 \left(\left[\frac{x}{L}\right] - \left[\frac{x}{L}\right]^2 \right)^{\frac{4}{3}}} \mathrm{d}x \tag{3.487}$$

$$= \frac{243 q_0^2 L^4}{199 \times 4 \times 4^{4/3} E a_0^4} \underbrace{\int_0^L \left(\left[\frac{x}{L}\right] - \left[\frac{x}{L}\right]^2 \right)^{\frac{2}{3}} \mathrm{d}x}_{0.2933414631064508\, L} . \tag{3.488}$$

In summary, one obtains for the optimized cross section:

$$a(x) = 4^{\frac{1}{3}} a_0 \left(- \left[\frac{x}{L}\right]^2 + \left[\frac{x}{L}\right]^1 \right)^{\frac{1}{3}}, \tag{3.489}$$

$$a_0 = \left(\frac{243}{796} \times \frac{q_0 L^2}{R_{\mathrm{p}0.2}} \right)^{\frac{1}{3}}, \tag{3.490}$$

$$V = \frac{8}{9} \times 4^{\frac{2}{3}} \times 0.293342 a_0^2 L, \tag{3.491}$$

$$\approx 0.657 \times a_0^2 L, \tag{3.492}$$

$$M \approx \frac{199}{54} \times \frac{1}{4^{\frac{2}{3}} \times 0.293342} \times \frac{a_0 R_{\mathrm{p}0.2}}{\varrho g L^2}, \tag{3.493}$$

$$\approx 4.985431 \times \frac{a_0 R_{\mathrm{p}0.2}}{\varrho g L^2}, \tag{3.494}$$

$$\Pi = \frac{199 \times 0.293342}{4^{\frac{1}{3}} \times 243} \times \frac{L a_0^2 R_{\mathrm{p}0.2}^2}{E}, \tag{3.495}$$

$$\approx 0.151333 \times \frac{L a_0^2 R_{\mathrm{p}0.2}^2}{E}, \tag{3.496}$$

$$SEA = \frac{199}{864} \times \frac{R_{\mathrm{p}0.2}^2}{\varrho E} \approx 0.230324 \times \frac{R_{\mathrm{p}0.2}^2}{\varrho E}. \tag{3.497}$$

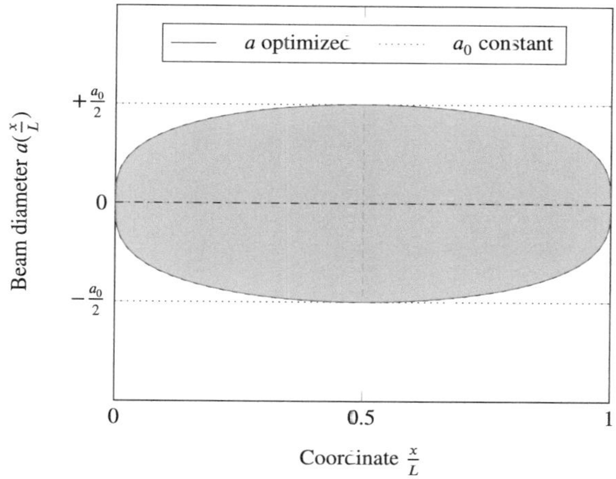

Fig. 3.25 Beam diameter along the beam axis

The optimized beam is shown in Fig. 3.25.

2.3.28 **Optimization of a beam with I-profile along the longitudinal axis under the influence of a constant distributed load**

- Moment distribution:

The bending moment distribution can be taken from the solution of Exercise 2.2.12:

$$M_y(x) = \frac{q_0}{2}(x^2 - Lx) = \frac{q_0 L^2}{2}\left(\left[\frac{x}{L}\right]^2 - \left[\frac{x}{L}\right]^1\right) < 0. \tag{3.498}$$

The maximum bending moment is obtained in the middle of the beam as:

$$M_{y,\max} = M_y\left(\tfrac{L}{2}\right) = -\frac{q_0 L^2}{8}. \tag{3.499}$$

- Geometric parameters of the cross section (I_y, A):

For an I-profile with external dimensions $a \times \alpha a$ and constant wall thickness $w = a/6$ the following results:

$$I_y = \frac{1}{12}a(\alpha a)^3 - 2 \times \frac{1}{12}\left(\frac{a}{2} - \frac{w}{2}\right)(\alpha a - 2w)^3 \tag{3.500}$$

$$= \frac{1}{12}\alpha^3 a^4 - \frac{5}{72}\left(\frac{3\alpha - 1}{3}\right)^3 a^4 \tag{3.501}$$

$$A = \alpha a^2 - 2\left(\frac{a}{2} - \frac{w}{2}\right)(\alpha a - 2w) \tag{3.502}$$

$$= \alpha a^2 - \frac{5}{6}\left(\frac{3\alpha - 1}{3}\right)a^2. \tag{3.503}$$

► Case $\alpha = 2$:

• Reference beam with constant cross section:

The evaluation of the stress criterion in the reference cross section at the location $x = L/2$, i.e.

$$\sigma_{x,\max} = \frac{M_{y,\max}}{I_y} \times a_0 = \frac{\frac{q_0 L^2}{8}}{\frac{671 a_0^4}{1944}} \times a_0 \stackrel{!}{=} R_{\text{p0.2}}, \tag{3.504}$$

results the expression for the distributed load as:

$$q_0 = \frac{671}{243} \times \frac{a_0^3 R_{\text{p0.2}}}{L^2}. \tag{3.505}$$

The definition equation of the lightweight index (see Eq. $(3.1)_{\text{TB}}$) results in:

$$M = \frac{F_0}{F_{\text{G}}}. \tag{3.506}$$

This gives with the approximation $F_0 \approx q_0 L$:.

$$M_0 \approx \frac{q_0 L}{\varrho g A_0 L} = \frac{q_0 L}{\varrho g L \frac{11}{18} a_0^2} = \frac{122}{27} \times \frac{a_0 R_{\text{p0.2}}}{\varrho g L^2}. \tag{3.507}$$

The total strain energy is calculated according to Eq. $(3.32)_{\text{TB}}$ for a bending load case as follows:

$$\Pi = \int_0^L \frac{M_y(x)^2}{2EI_y} \mathrm{d}x. \tag{3.508}$$

For our example, the total strain energy for the reference cross section is obtained by integration as:

$$\Pi_0 = \frac{1}{2E} \int_0^L \frac{\frac{q_0^2 L^4}{4}\left(\left[\frac{x}{L}\right] - \left[\frac{x}{L}\right]^2\right)^2}{\frac{671}{1944} a_0^4} \mathrm{d}x, \tag{3.509}$$

$$= \frac{1944 q_0^2 L^4}{8 \times 671 \times 4 E a_0^4} \underbrace{\int_0^L \left(\left[\frac{x}{L}\right]^2 - 2\left[\frac{x}{L}\right]^3 + \left[\frac{x}{L}\right]^4 \right) \mathrm{d}x}_{\frac{L}{30}}, \tag{3.510}$$

$$= \frac{1944 q_0^2 L^5}{8 \times 30 \times 671 E a_0^4} = \frac{671}{7290} \times \frac{L a_0^2 R_{\mathrm{p0.2}}^2}{E}. \tag{3.511}$$

In summary, one obtains for the reference beam:

$$A_0 = \frac{11 a_0^2}{18}, \tag{3.512}$$

$$I_0 = \frac{671 a_0^4}{1944}, \tag{3.513}$$

$$q_0 = \frac{671 a_0^3 R_{\mathrm{p0.2}}}{243 L^2}, \tag{3.514}$$

$$M_0 = \frac{122}{27} \times \frac{a_0 R_{\mathrm{p0.2}}}{\varrho g L^2}, \tag{3.515}$$

$$\approx 4.518519 \times \frac{a_0 R_{\mathrm{p0.2}}}{\varrho g L^2}, \tag{3.516}$$

$$\Pi_0 = \frac{671}{7290} \times \frac{L a_0^2 R_{\mathrm{p0.2}}^2}{E}, \tag{3.517}$$

$$\approx 0.0920439 \times \frac{L a_0^2 R_{\mathrm{p0.2}}^2}{E}, \tag{3.518}$$

$$SEA_0 = \frac{61}{405} \times \frac{R_{\mathrm{p0.2}}^2}{\varrho E}, \tag{3.519}$$

$$\approx 0.150617 \times \frac{R_{\mathrm{p0.2}}^2}{\varrho E}. \tag{3.520}$$

- Beam with optimized cross section:

Using the stress criterion

$$\sigma_{x,\max} = \frac{M_y(x)}{I_y(x)} \times a(x) = \frac{\frac{q_0 L^2}{2}\left(\left[\frac{x}{L}\right] - \left[\frac{x}{L}\right]^2\right)}{\frac{671 a(x)^4}{1944}} \times a(x) \overset{!}{=} R_{\mathrm{p0.2}}, \tag{3.521}$$

results

$$a^3(x) = \frac{1944 q_0 L^2}{2 \times 671 R_{\mathrm{p0.2}}} \times \left(\left[\frac{x}{L}\right] - \left[\frac{x}{L}\right]^2 \right). \tag{3.522}$$

The evaluation of the stress criterion in the reference cross section at the location $x = L/2$ provides the reference dimension (see Eq. (3.505)):

$$a_0^3 = \frac{1944 q_0 L^2}{8 \times 671 R_{\mathrm{p}0.2}}. \tag{3.523}$$

With this result we get:

$$a^3(x) = 4 \times \underbrace{\frac{1944 q_0 L^2}{8 \times 671 R_{\mathrm{p}0.2}}}_{a_0^3} \times \left(\left[\frac{x}{L}\right] - \left[\frac{x}{L}\right]^2 \right). \tag{3.524}$$

The volume is determined by evaluating the integral:

$$V = \int_0^L A(x)\,\mathrm{d}x = \frac{11 \times 4^{\frac{2}{3}} a_0^2}{18} \underbrace{\int_0^L \left(\left[\frac{x}{L}\right] - \left[\frac{x}{L}\right]^2 \right)^{\frac{2}{3}} \mathrm{d}x}_{0.2933414631064508\,L}. \tag{3.525}$$

The lightweight index is again approximately:

$$M \approx \frac{q_0 L}{\varrho g V} = \frac{q_0 L}{\varrho g \times \frac{11}{18} \times 4^{2/3} a_0^2 \times 0.293341\,L}. \tag{3.526}$$

The total strain energy is:

$$\Pi = \frac{1}{2E} \int_0^L \frac{\frac{q_0^2 L^4}{4} \left(\left[\frac{x}{L}\right] - \left[\frac{x}{L}\right]^2 \right)^2}{\frac{671}{1944} 4^{4/3} a_0^4 \left(\left[\frac{x}{L}\right] - \left[\frac{x}{L}\right]^2 \right)^{\frac{4}{3}}} \mathrm{d}x \tag{3.527}$$

$$= \frac{243 q_0^2 L^4}{671 \times 4^{4/3} E a_0^4} \underbrace{\int_0^L \left(\left[\frac{x}{L}\right] - \left[\frac{x}{L}\right]^2 \right)^{\frac{2}{3}} \mathrm{d}x}_{0.2933414631064508\,L}. \tag{3.528}$$

In summary, one obtains for the optimized cross section:

$$a(x) = 4^{\frac{1}{3}} a_0 \left(- \left[\frac{x}{L} \right]^2 + \left[\frac{x}{L} \right]^1 \right)^{\frac{1}{3}}, \tag{3.529}$$

$$a_0 = \left(\frac{1}{4} \times \frac{972}{671} \times \frac{q_0 L^2}{R_{\text{p0.2}}} \right)^{\frac{1}{3}}, \tag{3.530}$$

$$V \approx \frac{11}{18} \times 4^{\frac{2}{3}} \times 0.293342 a_0^2 L, \tag{3.531}$$

$$\approx 0.452 \times a_0^2 L, \tag{3.532}$$

$$M \approx \frac{122}{27} \times \frac{1}{4^{\frac{2}{3}} \times 0.293342} \times \frac{a_0 R_{\text{p0.2}}}{\varrho g L^2}, \tag{3.533}$$

$$\approx 6.112917 \times \frac{a_0 R_{\text{p0.2}}}{\varrho g L^2}, \tag{3.534}$$

$$\Pi \approx \frac{671 \times 0.293342}{243 \times 4^{\frac{4}{3}}} \times \frac{L a_0^2 R_{\text{p0.2}}^2}{E}, \tag{3.535}$$

$$\approx 0.127569 \times \frac{L a_0^2 R_{\text{p0.2}}^2}{E}, \tag{3.536}$$

$$SEA \approx 0.282407 \times \frac{R_{\text{p0.2}}^2}{\varrho E}. \tag{3.537}$$

The optimized beam is shown in Fig. 3.26.

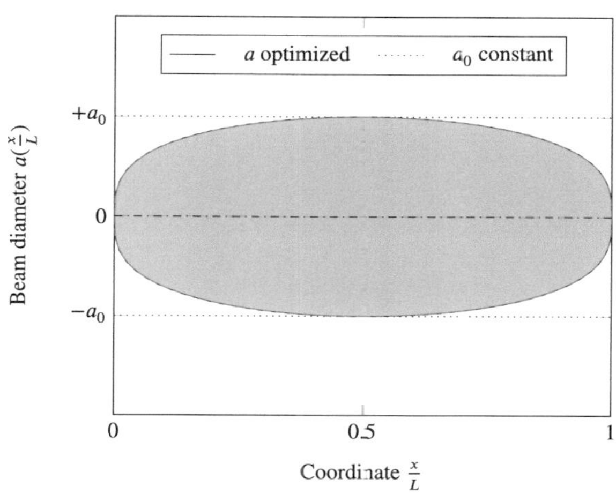

Fig. 3.26 Beam diameter along the beam axis

▶ Case $\alpha = 3$:

• Reference beam with constant cross section:

The evaluation of the stress criterion in the reference cross section at the location $x = L/2$, i.e.

$$\sigma_{x,\max} = \frac{M_{y,\max}}{I_y} \times a_0 = \frac{\frac{q_0 L^2}{8}}{\frac{907 a_0^4}{972}} \times \frac{3a_0}{2} \overset{!}{=} R_{\text{p0.2}}, \tag{3.538}$$

results the expression for the distributed load as:

$$q_0 = \frac{3628}{729} \times \frac{a_0^3 R_{\text{p0.2}}}{L^2}. \tag{3.539}$$

The lightweight index is calculated with the approximation $F_0 \approx q_0 L$ to:

$$M_0 \approx \frac{q_0 L}{\varrho g A_0 L} = \frac{q_0 L}{\varrho g L \frac{7}{9} a_0^2} = \frac{3628}{567} \times \frac{a_0 R_{\text{p0.2}}}{\varrho g L^2}. \tag{3.540}$$

For our example, the total strain energy for the reference cross section is obtained by integration as:

$$\Pi_0 = \frac{1}{2E} \int_0^L \frac{\frac{q_0^2 L^4}{4} \left(\left[\frac{x}{L}\right] - \left[\frac{x}{L}\right]^2 \right)^2}{\frac{907}{972} a_0^4} \mathrm{d}x, \tag{3.541}$$

$$= \frac{972 q_0^2 L^4}{8 \times 907 E a_0^4} \underbrace{\int_0^L \left(\left[\frac{x}{L}\right]^2 - 2\left[\frac{x}{L}\right]^3 + \left[\frac{x}{L}\right]^4 \right) \mathrm{d}x}_{\frac{L}{30}}, \tag{3.542}$$

$$= \frac{972 q_0^2 L^5}{8 \times 30 \times 907 E a_0^4} = \frac{3628}{32805} \times \frac{L a_0^2 R_{\text{p0.2}}^2}{E}. \tag{3.543}$$

In summary, one obtains for the reference beam:

$$A_0 = \frac{7a_0^2}{9}, \tag{3.544}$$

$$I_0 = \frac{907 a_0^4}{972}, \tag{3.545}$$

$$q_0 = \frac{3628 a_0^3 R_{\text{p0.2}}}{729 L^2}, \tag{3.546}$$

$$M_0 = \frac{3628}{567} \times \frac{a_0 R_{\text{p0.2}}}{\varrho g L^2}, \tag{3.547}$$

$$\approx 6.398589 \times \frac{a_0 R_{p0.2}}{\varrho g L^2}, \tag{3.548}$$

$$\Pi_0 = \frac{3628}{32805} \times \frac{L a_0^2 R_{p0.2}^2}{E}, \tag{3.549}$$

$$\approx 0.110593 \times \frac{L a_0^2 R_{p0.2}^2}{E}, \tag{3.550}$$

$$SEA_0 = \frac{3628}{25515} \times \frac{R_{p0.2}^2}{\varrho E}, \tag{3.551}$$

$$\approx 0.142191 \times \frac{R_{p0.2}^2}{\varrho E}. \tag{3.552}$$

- Beam with optimized cross section:

Using the stress criterion

$$\sigma_{x,\max} = \frac{M_y(x)}{I_y(x)} \times a(x) = \frac{\frac{q_0 L^2}{2}\left(\left[\frac{x}{L}\right] - \left[\frac{x}{L}\right]^2\right)}{\frac{907 a(x)^4}{972}} \times \frac{3a(x)}{2} \stackrel{!}{=} R_{p0.2}, \tag{3.553}$$

results

$$a^3(x) = \frac{729 q_0 L^2}{907 R_{p0.2}} \times \left(\left[\frac{x}{L}\right] - \left[\frac{x}{L}\right]^2\right). \tag{3.554}$$

The evaluation of the stress criterion in the reference cross section at the location $x = L/2$ provides the reference dimension (see Eq. (3.539)):

$$a_0^3 = \frac{729 q_0 L^2}{3628 R_{p0.2}}. \tag{3.555}$$

With this result we get:

$$a^3(x) = 4 \times \underbrace{\frac{729 q_0 L^2}{4 \times 907 R_{p0.2}}}_{a_0^3} \times \left(\left[\frac{x}{L}\right] - \left[\frac{x}{L}\right]^2\right). \tag{3.556}$$

The volume is determined by evaluating the integral:

$$V = \int_0^L A(x)\,dx = \frac{7 \times 4^{\frac{2}{3}} a_0^2}{9} \underbrace{\int_0^L \left(\left[\frac{x}{L}\right] - \left[\frac{x}{L}\right]^2\right)^{\frac{2}{3}} dx}_{0.2933414631064508\,L}. \tag{3.557}$$

The lightweight index is again approximately:

$$M \approx \frac{q_0 L}{\varrho g V} = \frac{q_0 L}{\varrho g \times \frac{9}{9} \times 4^{2/3} a_0^2 \times 0.293341\, L}. \tag{3.558}$$

The total strain energy is:

$$\Pi = \frac{1}{2E} \int\limits_0^L \frac{\frac{q_0^2 L^4}{4} \left(\left[\frac{x}{L}\right] - \left[\frac{x}{L}\right]^2 \right)^2}{\frac{907}{972} 4^{4/3} a_0^4 \left(\left[\frac{x}{L}\right] - \left[\frac{x}{L}\right]^2 \right)^{\frac{4}{3}}} \mathrm{d}x \tag{3.559}$$

$$= \frac{972 q_0^2 L^4}{907 \times 8 \times 4^{4/3} E a_0^4} \underbrace{\int\limits_0^L \left(\left[\frac{x}{L}\right] - \left[\frac{x}{L}\right]^2 \right)^{\frac{2}{3}} \mathrm{d}x}_{0.2933414631064508\, L} . \tag{3.560}$$

In summary, one obtains for the optimized cross section:

$$a(x) = 4^{\frac{1}{3}} a_0 \left(- \left[\frac{x}{L}\right]^2 + \left[\frac{x}{L}\right]^1 \right)^{\frac{1}{3}}, \tag{3.561}$$

$$a_0 = \left(\frac{1}{4} \times \frac{729}{907} \times \frac{q_0 L^2}{R_{\mathrm{p}0.2}} \right)^{\frac{1}{3}}, \tag{3.562}$$

$$V \approx \frac{7}{9} \times 4^{\frac{2}{3}} \times 0.293342 a_0^2 L, \tag{3.563}$$

$$\approx 0.575 \times a_0^2 L, \tag{3.564}$$

$$M \approx \frac{3628}{567} \times \frac{1}{4^{\frac{2}{3}} \times 0.293342} \times \frac{a_0 R_{\mathrm{p}0.2}}{\varrho g L^2}, \tag{3.565}$$

$$\approx 8.656387 \times \frac{a_0 R_{\mathrm{p}0.2}}{\varrho g L^2}, \tag{3.566}$$

$$\Pi \approx \frac{7256 \times 0.293342}{2187 \times 4^{\frac{4}{3}}} \times \frac{L a_0^2 R_{\mathrm{p}0.2}^2}{E}, \tag{3.567}$$

$$\approx 0.153277 \times \frac{L a_0^2 R_{\mathrm{p}0.2}^2}{E}, \tag{3.568}$$

$$SEA \approx 0.266608 \times \frac{R_{\mathrm{p}0.2}^2}{\varrho E}. \tag{3.569}$$

The optimized beam is shown in Fig. 3.26.

2.3.29 Optimization of a beam with I-profile along the longitudinal axis and the height-to-width ratio under the influence of a constant distributed load

The basic procedure can be found in the solution to the Exercise 2.3.28. Depending on the height-to-width ratio, the following relationships arise:

$$I_y(x) = \left(\frac{\alpha^3}{12} - \frac{5}{72}\left(\frac{3\alpha-1}{3}\right)^3\right) a^4(x), \tag{3.570}$$

$$A(x) = \left(\alpha - \frac{5}{6}\left(\frac{3\alpha-1}{3}\right)\right) a^2(x) \tag{3.571}$$

$$a_0^3 = \frac{\alpha q_0 L^2}{16\left(\frac{\alpha^3}{12} - \frac{5}{72}\left(\frac{3\alpha-1}{3}\right)^3\right) R_{\text{p0.2}}}, \tag{3.572}$$

$$a(x) = 4^{\frac{1}{3}} a_0 \left(\left[\frac{x}{L}\right]^1 - \left[\frac{x}{L}\right]^2\right)^{\frac{1}{3}}, \tag{3.573}$$

$$V = \left(\alpha - \frac{5}{6}\left(\frac{3\alpha-1}{3}\right)\right) 4^{\frac{2}{3}} a_0^2 \times 0.293342L, \tag{3.574}$$

$$\Pi = \frac{0.293342 \times L^5 q_0^2}{8 \times 4^{4/3} E a_0^4 \left(\frac{\alpha^3}{12} - \frac{5}{72}\left(\frac{3\alpha-1}{3}\right)^3\right)} \tag{3.575}$$

$$= \frac{32 \times 0.293342}{4^{4/3}} \times \frac{\left(\frac{\alpha^3}{12} - \frac{5}{72}\left(\frac{3\alpha-1}{3}\right)^3\right)}{\alpha^2} \times \frac{R_{\text{p0.2}}^2 a_0^2 L}{E}, \tag{3.576}$$

$$SEA = 2 \times \frac{R_{\text{p0.2}}^2}{\varrho E} \times \frac{\frac{\alpha^3}{12} - \frac{5}{72}\left(\frac{3\alpha-1}{3}\right)^3}{\alpha^2\left(\alpha - \frac{5}{6}\left(\frac{3\alpha-1}{3}\right)\right)}. \tag{3.577}$$

The evaluation of the specific energy absorption SEA of the optimized cross section for different values of α is shown in Fig. 3.27.

2.3.30 Optimization of a beam along the longitudinal axis under the influence of two different single forces

- Internal reactions:

To determine the internal reactions, the beam under the influence of two different individual forces is cut free at an arbitrary location x to the left and then to the right of the left single force, see Fig. 3.28.

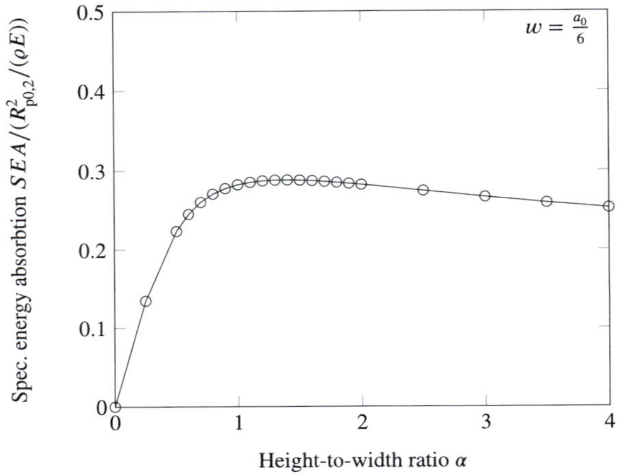

Fig. 3.27 Specific energy absorbtion for different values of α

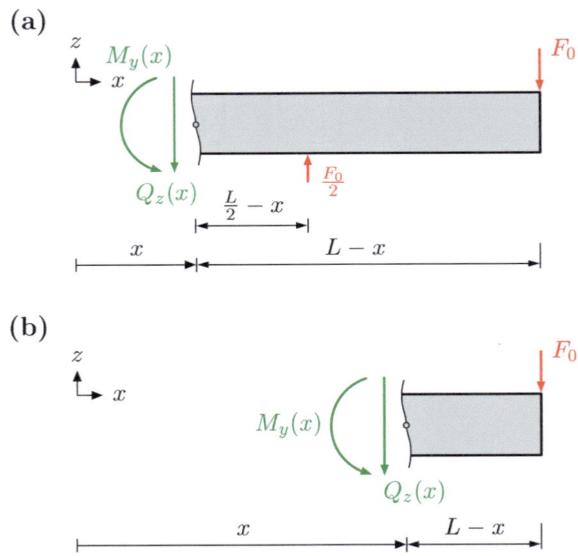

Fig. 3.28 Cutting free inside the beam under the influence of two different single forces: **a** Section 1 and **b** Section 2

The moment equilibrium around the reference point at the location x in the Section 1 ($0 \leq x \leq \frac{L}{2}$) results in the distribution of the internal bending moment, see Fig. 3.28a:

$$\overset{\curvearrowright}{\sum} M_y = 0: \quad +F_0(L-x) - \frac{F_0}{2}\left(\frac{L}{2} - x\right) - M_y(x) = 0, \tag{3.578}$$

$$M_y(x) = \frac{3F_0L}{4}\left(1 - \frac{2x}{3L}\right) \quad \text{for} \quad 0 \leq x \leq \frac{L}{2}. \tag{3.579}$$

The moment equilibrium around the reference point at the location x in the Section 2 ($\frac{L}{2} \leq x \leq L$) results in the distribution of the internal bending moment, see Fig. 3.28b:

$$\overset{\curvearrowright}{\sum} M_y = 0: \quad +F_0(L-x) - M_y(x) = 0, \tag{3.580}$$

$$M_y(x) = F_0(L-x) \quad \text{for} \quad \frac{L}{2} \leq x \leq L. \tag{3.581}$$

In summary, the two moment distributions can be given as follows:

$$M_{y,\mathrm{I}}(x) = \frac{3F_0L}{4}\left(1 - \frac{2x}{3L}\right) \quad \text{for} \quad 0 \leq x \leq \frac{L}{2}, \tag{3.582}$$

$$M_{y,\mathrm{II}}(x) = F_0L\left(1 - \frac{x}{L}\right) \quad \text{for} \quad \frac{L}{2} \leq x \leq L. \tag{3.583}$$

- Geometric parameters of the cross section (I_y, A):

For a box profile with external dimensions $d \times d$ and constant wall thickness $s = d/5$ we get (see Exercise 2.1.6):

$$I_y = \frac{1}{12}\left(d_0^2 + \left(d_0 - \frac{2d_0}{5}\right)^2\right)\left(d_0 - \frac{3d_0}{5}\right)\left(d_0 + \frac{3d_0}{5}\right) \tag{3.584}$$

$$= \frac{136}{1875}d_0^5, \tag{3.585}$$

$$A = \left(d_0 - \frac{3d_0}{5}\right)\left(d_0 + \frac{3d_0}{5}\right) \tag{3.586}$$

$$= \frac{16}{25}d_0^2. \tag{3.587}$$

- Optimized diameter:

Using the stress criterion in the Section 1 ($0 \leq x \leq \frac{L}{2}$)

$$\sigma_{x,\max} = \frac{M_y(x)}{I_y(x)} \times \frac{d(x)}{2} = \frac{\frac{3F_0L}{4}\left(1 - \frac{2x}{3L}\right)}{\frac{136}{1875}d^4(x)} \times \frac{d(x)}{2} \overset{!}{=} R_{\mathrm{p}0.2}, \tag{3.588}$$

results

$$d^3(x) = \frac{3 \times 1875 F_0 L}{8 \times 136 R_{\mathrm{p}0.2}} \times \left(1 - \frac{2x}{3L}\right). \tag{3.589}$$

The evaluation of the stress criterion in the reference cross section, i.e. for $x = 0$, results in:

$$\sigma_{x,\max} = \frac{M_y(x=0)}{I_y} \times \frac{d_0}{2} = \frac{\frac{3F_0L}{4}}{\frac{136}{1875}d_0^4} \times \frac{d_0}{2} \overset{!}{=} R_{\text{p0.2}}, \tag{3.590}$$

or

$$d_0^3 = \frac{3 \times 1875 F_0 L}{8 \times 136 R_{\text{p0.2}}}. \tag{3.591}$$

With this result, the optimized cross section in the first section is calculated as:

$$d^3(x) = d_0^3 \times \left(1 - \frac{2x}{3L}\right). \tag{3.592}$$

Using the stress criterion in the Section 2 ($\frac{L}{2} \leq x \leq L$)

$$\sigma_{x,\max} = \frac{M_y(x)}{I_y^(x)} \times \frac{d(x)}{2} = \frac{F_0L\left(1-\frac{x}{L}\right)}{\frac{136}{1875}d^4(x)} \times \frac{d(x)}{2} \overset{!}{=} R_{\text{p0.2}}, \tag{3.593}$$

results

$$d^3(x) = \frac{4}{3} \times \frac{3 \times 1875 F_0 L}{8 \times 136 R_{\text{p0.2}}}\left(1 - \frac{x}{L}\right) = \frac{4}{3} \times d_0^3\left(1 - \frac{x}{L}\right). \tag{3.594}$$

In summary, the following distribution results for both sections (see also Fig. 3.29):

$$d_{\text{I}}(x) = d_0 \times \sqrt[3]{1 - \frac{2x}{3L}} \quad \text{for} \quad 0 \leq x \leq \frac{L}{2}, \tag{3.595}$$

$$d_{\text{II}}(x) = \sqrt[3]{\frac{4}{3}} \times d_0 \times \sqrt[3]{1 - \frac{x}{L}} \quad \text{for} \quad \frac{L}{2} \leq x \leq L, \tag{3.596}$$

where the reference diameter at the clamping point is as follows:

$$d_0 = \sqrt[3]{\frac{\frac{3F_0L}{4}}{\frac{272}{1875}R_{\text{p0.2}}}}. \tag{3.597}$$

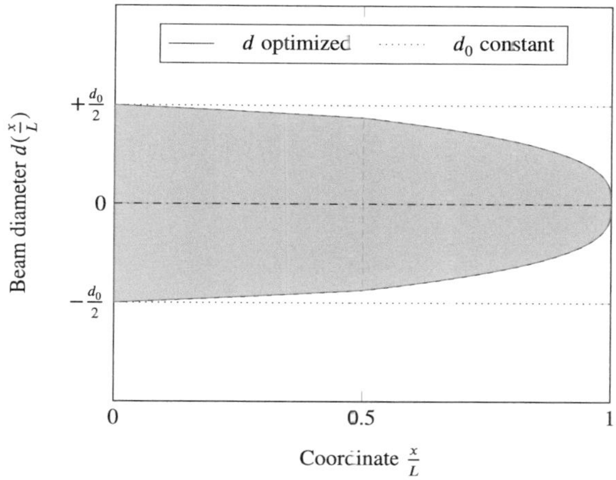

Fig. 3.29 Beam diameter (outer) along the beam axis

- Total strain energy:

$$\Pi = \int_0^L \frac{(M_y(x))^2}{2EI}\mathrm{d}x = \int_0^{\frac{L}{2}} \frac{M_{y,\mathrm{I}}(x)^2}{2EI_\mathrm{I}}\mathrm{d}x + \int_{\frac{L}{2}}^{L} \frac{(M_{y,\mathrm{II}}(x))^2}{2EI_\mathrm{II}}\mathrm{d}x \tag{3.598}$$

$$= \frac{1}{2E}\int_0^{\frac{L}{2}} \frac{\frac{9F_0^2L^2}{16}\left(1-\frac{2x}{3L}\right)^2}{\frac{136}{1875}d_0^4\left(1-\frac{2x}{3L}\right)^{4/3}}\mathrm{d}x + \frac{1}{2E}\int_{\frac{L}{2}}^{L} \frac{F_0^2L^2\left(1-\frac{x}{L}\right)^2}{\frac{136}{1875}d_0^4\left(\frac{4}{3}\right)^{4/3}\left(1-\frac{x}{L}\right)^{4/3}}\mathrm{d}x \tag{3.599}$$

$$= \frac{16875 \times F_0^2L^2\left(\frac{9L}{10} - \frac{2^{\frac{2}{3}}3^{\frac{1}{3}}L}{5}\right)}{4352 \times Ed_0^4} + \frac{125 \times 3^{\frac{10}{3}}F_0^2L^3}{17 \times 2^{\frac{5}{3}}4^{\frac{10}{3}}Ed_0^4} \tag{3.600}$$

$$\approx 2.602043 \times \frac{F_0^2L^3}{Ed_0^4}. \tag{3.601}$$

- Total mass:

$$m = \int_0^L \varrho A(x)\mathrm{d}x = \int_0^{\frac{L}{2}} \varrho A_\mathrm{I}(x)\mathrm{d}x + \int_{\frac{L}{2}}^{L} \varrho A_\mathrm{II}(x)\mathrm{d}x \tag{3.602}$$

$$= \varrho\frac{16}{25}d_0^2\int_0^{\frac{L}{2}}\left(1-\frac{2x}{3L}\right)^{\frac{2}{3}}\mathrm{d}x + \varrho\frac{16}{25}\left(\frac{4}{3}\right)^{\frac{2}{3}}d_0^2\int_{\frac{L}{2}}^{L}\left(1-\frac{x}{L}\right)^{\frac{2}{3}}\mathrm{d}x \tag{3.603}$$

$$= \frac{16}{25} \times \left(\frac{9L}{10} - \frac{2^{\frac{2}{3}} 3^{\frac{1}{3}} L}{5} \right) d_0^2 \varrho + \frac{3^{\frac{1}{3}} 4^{\frac{8}{3}} L d_0^2 \varrho}{125 \times 2^{\frac{5}{3}}} \tag{3.604}$$

$$\approx 0.429477 \times L \, {d_0}^2 \varrho. \tag{3.605}$$

- Specific energy absorption:

$$SEA = \frac{\Pi}{m} = 6.058637 \times \frac{F_0^2 L^2}{E d_0^6 \varrho}. \tag{3.606}$$

2.3.31 **Optimization of a beam along the longitudinal axis under the influence of a single force and a single moment**

- Internal reactions:

To determine the internal reactions, the beam under the influence of a single force and a single moment is cut free at an arbitrary location x inside the beam, see Fig. 3.30.

The moment equilibrium around the reference point at the location x results in the distribution of the internal bending moment:

$$\overset{\curvearrowright}{\sum} M_y = 0: \quad +F_0(L - x) + M_0 - M_y(x) = 0, \tag{3.607}$$

$$M_y(x) = F_0 L \left(1 - \frac{x}{L}\right) + M_0 = F_0 L \left(1 - \frac{x}{L}\right) + \frac{F_0 L}{6} \tag{3.608}$$

$$= F_0 L \left(\frac{7}{6} - \frac{x}{L}\right). \tag{3.609}$$

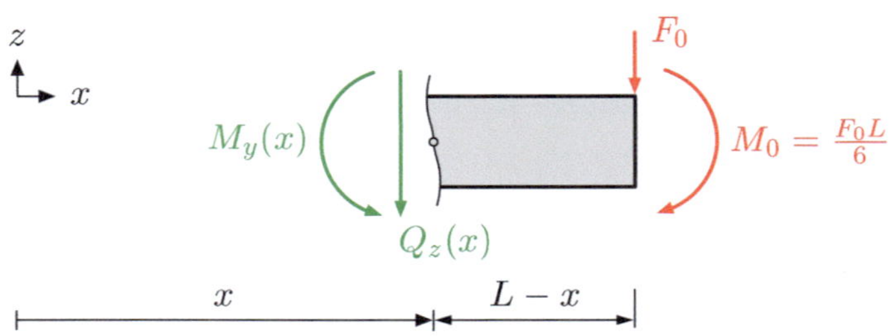

Fig. 3.30 Cutting free inside the beam under the influence of a single force and a single moment

• Geometric properties of the cross section (I_y, A):

The geometric properties can be taken from the solution to the Exercise 2.3.29:

$$I_y(x) = \left(\frac{\alpha^3}{12} - \frac{5}{72} \left(\frac{3\alpha - 1}{3} \right)^3 \right) a^4(x), \tag{3.610}$$

$$A(x) = \left(\alpha - \frac{5}{6} \left(\frac{3\alpha - 1}{3} \right) \right) a^2(x). \tag{3.611}$$

Setting $\alpha = 1$, one gets:

$$I(x) = \frac{61a(x)^4}{972}, \tag{3.612}$$

$$A(x) = \frac{4a(x)^2}{9}. \tag{3.613}$$

• Beam contour $a = a(x)$ along the beam axis:

Using the stress criterion

$$\sigma_{x,\max} = \frac{M_y(x)}{I_y(x)} \times \frac{a(x)}{2} = \frac{F_0 L \left(\frac{7}{6} - \frac{x}{L} \right)}{\frac{61a(x)^4}{972}} \times \frac{a(x)}{2} \overset{!}{=} R_{p0.2}, \tag{3.614}$$

results

$$a^3(x) = \frac{972 F_0 L}{122 R_{p0.2}} \times \left(\frac{7}{3} - \frac{x}{L} \right). \tag{3.615}$$

The evaluation of the stress criterion in the reference cross section at the location $x = 0$ provides the reference dimension to:

$$a_0^3 = \frac{7 \times 81 F_0 L}{61 R_{p0.2}}. \tag{3.616}$$

With this result we get:

$$a^3(x) = \frac{6}{7} \times \underbrace{\frac{7 \times 972 F_0 L}{6 \times 122 R_{p0.2}}}_{a_0^3} \times \left(\frac{7}{6} - \frac{x}{L} \right), \tag{3.617}$$

respectively (see Fig. 3.31)

$$a(x) = \left(\frac{6}{7} \right)^{\frac{1}{3}} a_0 \left(\frac{7}{6} - \frac{x}{L} \right)^{\frac{1}{3}} \approx 0.9499 \times a_0 \left(\frac{7}{6} - \frac{x}{L} \right)^{\frac{1}{3}}, \tag{3.618}$$

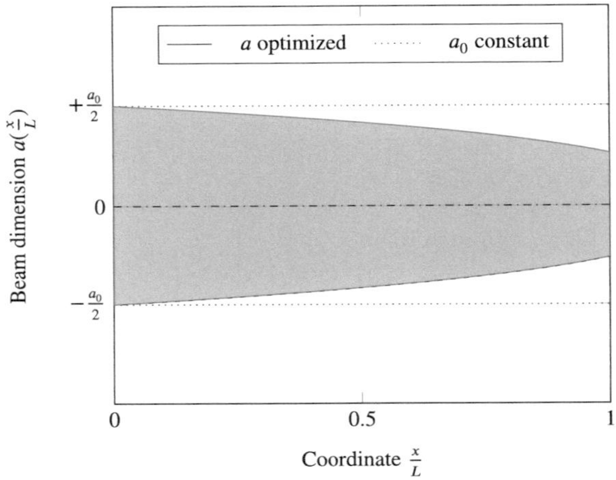

Fig. 3.31 Beam dimension along the beam axis

where the following dimension results for the reference cross section:

$$a_0 = \left(\frac{567 F_0 L}{61 R_{p0.2}}\right)^{\frac{1}{3}}. \tag{3.619}$$

• Constant cross section:

The total strain energy is calculated according to Eq. (3.32)$_{\text{TB}}$ for a bending load case as follows:

$$\Pi = \int_0^L \frac{M_y(x)^2}{2EI_y} \mathrm{d}x. \tag{3.620}$$

For our example, the total strain energy for the reference cross section is obtained by integration as:

$$\Pi_0 = \frac{1}{2E} \int_0^L \frac{F_0^2 L^2 \left(\frac{7}{6} - \frac{x}{L}\right)^2}{\frac{61}{972} a_0^4} \mathrm{d}x, \tag{3.621}$$

$$= \frac{972 F_0^2 L^2}{122 E a_0^4} \underbrace{\int_0^L \left(\frac{7}{6} - \frac{x}{L}\right)^2 \mathrm{d}x}_{\frac{19L}{36}}, \tag{3.622}$$

$$= \frac{513 F_0^2 L^3}{122 E a_0^4} \approx 4.2049180 \times \frac{F_0^2 L^3}{E a_0^4} \tag{3.623}$$

$$= \frac{1159}{23814} \times \frac{L a_0^2 R_{p0.2}^2}{E} \approx 0.0486689 \times \frac{L a_0^2 R_{p0.2}^2}{E}. \tag{3.624}$$

The specific energy absorption is generally given by Eq. (3.27)$_{\text{TB}}$ as:

$$SEA = \frac{\Pi}{m}, \tag{3.625}$$

or taking into account the expressions for the total strain energy (see Eq. (3.624)) and mass ($m = \varrho L \frac{4}{9} a_0^2$):

$$SEA_0 = \frac{1159}{10584} \times \frac{R_{\text{p0.2}}^2}{E\varrho} \approx 0.109505 \times \frac{R_{\text{p0.2}}^2}{E\varrho}. \tag{3.626}$$

• Optimized cross section::

The total strain energy for the optimized cross section is obtained by integration as:

$$\Pi = \frac{1}{2E} \int_0^L \frac{F_0^2 L^2 \left(\frac{7}{6} - \frac{x}{L}\right)^2}{\frac{61}{972} \left(\frac{6}{7}\right)^{\frac{4}{3}} a_0^4 \left(\frac{7}{6} - \frac{x}{L}\right)^{\frac{4}{3}}} \mathrm{d}x \tag{3.627}$$

$$= \frac{972}{122 \left(\frac{6}{7}\right)^{4/3}} \times \frac{F_0^2 L^2}{a_0^4} \underbrace{\int_0^L \left(\frac{7}{6} - \frac{x}{L}\right)^{\frac{2}{3}} \mathrm{d}x}_{0.852346\,L} \tag{3.628}$$

$$\approx 7.294636 \times \frac{F_0^2 L^3}{E a_0^4} \approx 0.084430 \times \frac{R_{\text{p0.2}}^2 L a_0^2}{E}. \tag{3.629}$$

The volume is determined by evaluating the integral:

$$V = \int_0^L A(x)\,\mathrm{d}x = \int_0^L \frac{4}{9} a^2(x)\,\mathrm{d}x \approx 0.298966 a_0^2 L. \tag{3.630}$$

With these values, the specific energy absorption is:

$$SEA \approx 0.282407 \times \frac{R_{\text{p0.2}}^2}{E\varrho}. \tag{3.631}$$

• Ratio:

$$\frac{SEA}{SEA_0} = \frac{0.282407}{0.109505} \approx 2.5789. \tag{3.632}$$

2.3.32 **Optimization of a cantilever beam along the longitudinal axis under the influence of a constant distributed load**

• Moment distribution:

The bending moment distribution can be taken from the solution of Exercise 2.1.2b:

$$M_y(x) = \frac{q_0 L^2}{2}\left(1 - \frac{x}{L}\right)^2. \tag{3.633}$$

The maximum bending moment is obtained at the clamping point of the beam as:

$$M_{y,\max} = M_y\,(x = 0) = \frac{q_0 L^2}{2}. \tag{3.634}$$

• Geometric properties of the cross section (I_y, A):

For a box profile with external dimensions $a \times 2a$ and constant wall thickness $d = a/6$, the Exercise 2.3.27 provides:

$$I_y = \frac{1}{12}a(2a)^3 - \frac{1}{12}(a - 2d)(2a - 2d)^3 = \frac{199}{486}a^4, \tag{3.635}$$

$$A = 2a^2 - (a - 2d)(2a - 2d) = \frac{8}{9}a^2. \tag{3.636}$$

• Constant cross section:

The total strain energy is calculated according to Eq. $(3.32)_{\mathrm{TB}}$ for a bending load case as follows:

$$\Pi = \int_0^L \frac{M_y(x)^2}{2EI_y}\mathrm{d}x. \tag{3.637}$$

For our example, the total strain energy for the reference cross section is obtained by integration as:

$$\Pi_0 = \frac{1}{2E}\int_0^L \frac{\frac{q_0^2 L^4}{4}\left(1 - \frac{x}{L}\right)^4}{\frac{199}{486}a_0^4}\mathrm{d}x, \tag{3.638}$$

$$= \frac{486 q_0^2 L^4}{8 \times 199 E a_0^4}\underbrace{\int_0^L \left(1 - \frac{x}{L}\right)^4 \mathrm{d}x}_{\frac{L}{5}}, \tag{3.639}$$

$$= \frac{243 q_0^2 L^5}{3980 E a_0^4}. \tag{3.640}$$

The specific energy absorption is generally given by Eq. $(3.27)_{\text{TB}}$ as:

$$SEA = \frac{\Pi}{m}, \tag{3.641}$$

or taking into account the expressions for the total strain energy (see Eq. (3.640)) and mass ($m = \varrho L \frac{8}{9} a_0^2$):

$$SEA_0 = \frac{2187 q_0^2 L^4}{31840 E a_0 \varrho}. \tag{3.642}$$

In summary, one obtains for the reference beam:

$$A_0 = \frac{8 a_0{}^2}{9}, \tag{3.643}$$

$$I_0 = \frac{199 a_0{}^4}{486}, \tag{3.644}$$

$$m_0 = \frac{8\, L\, a_0{}^2 \varrho}{9}, \tag{3.645}$$

$$\Pi_0 = \frac{243 L^5\, q_0{}^2}{3980 E\, a_0{}^4}, \tag{3.646}$$

$$SEA_0 = \frac{2187 L^4\, q_0{}^2}{31840 E\, a_0{}^6 \varrho}. \tag{3.647}$$

• Optimized cross section:

Using the stress criterion

$$\sigma_{x,\max} = \frac{M_y(x)}{I_y(x)} \times a(x) = \frac{\frac{q_0 L^2}{2}\left(1 - \frac{x}{L}\right)^2}{\frac{199 a(x)^4}{486}} \times a(x) \stackrel{!}{=} R_{\text{p0.2}}, \tag{3.648}$$

results

$$a^3(x) = \frac{486 q_0 L^2}{398 R_{\text{p0.2}}} \times \left(1 - \frac{x}{L}\right)^2. \tag{3.649}$$

The evaluation of the stress criterion in the reference cross section at the location $x = 0$ provides the reference dimension to:

$$a_0^3 = \frac{486 q_0 L^2}{398 R_{\text{p0.2}}}. \tag{3.650}$$

With this result we get:

$$a^3(x) = a_0^3 \times \left(1 - \frac{x}{L}\right)^2. \tag{3.651}$$

or

$$a(x) = a_0 \times \left(1 - \frac{x}{L}\right)^{\frac{2}{3}}. \tag{3.652}$$

The volume is determined by evaluating the integral:

$$V = \int_0^L A(x)\,\mathrm{d}x = \int_0^L \frac{8}{9}a^2(x)\,\mathrm{d}x = \frac{8}{9}a_0^2 \underbrace{\int_0^L \left(1 - \frac{x}{L}\right)^{\frac{4}{3}}}_{\frac{3L}{7}} \tag{3.653}$$

$$= \frac{8}{21} \times a_0^2 L. \tag{3.654}$$

The total strain energy for the optimized beam is obtained by integration as:

$$\Pi = \frac{1}{2E}\int_0^L \frac{\frac{q_0^2 L^4}{4}\left(1 - \frac{x}{L}\right)^4}{\frac{199}{486}a_0^4\left(1 - \frac{x}{L}\right)^{\frac{8}{3}}}\mathrm{d}x \tag{3.655}$$

$$= \frac{486}{398} \times \frac{q_0^2 L^4}{E a_0^4} \underbrace{\int_0^L \left(1 - \frac{x}{L}\right)^{\frac{4}{3}} \mathrm{d}x}_{\frac{3L}{7}} \tag{3.656}$$

$$\frac{729}{5572} \times \frac{q_0^2\,L^5}{E a_0^4}. \tag{3.657}$$

In summary, one obtains for the optimized cross section (see Fig. 3.32):

$$a(x) = a_0\left(1 - \frac{x}{L}\right)^{\frac{2}{3}}, \tag{3.658}$$

$$A(x) = \frac{8}{9}\mathrm{a}(x)^2, \tag{3.659}$$

$$I(x) = \frac{199}{486}\mathrm{a}(x)^4, \tag{3.660}$$

$$m = \frac{8\,L\,{a_0}^2\varrho}{21}, \tag{3.661}$$

$$\Pi = \frac{729L^5\,{q_0}^2}{5572E\,{a_0}^4}, \tag{3.662}$$

$$SEA = \frac{2187L^4\,{q_0}^2}{6368E\,{a_0}^6\varrho}. \tag{3.663}$$

Fig. 3.32 Beam dimension along the beam axis

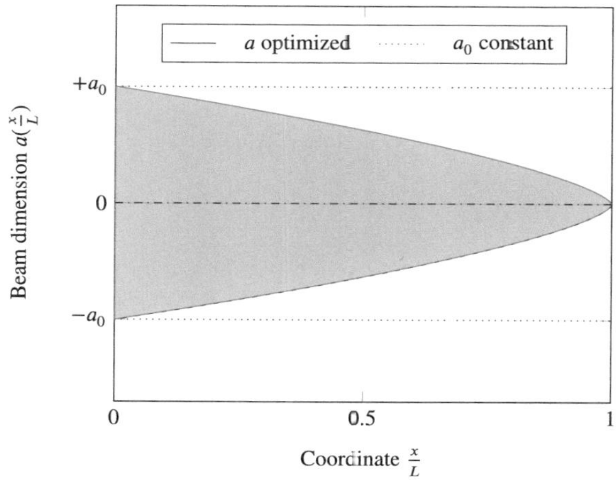

• Ratios:

$$\frac{m}{m_0} = \frac{3}{7}, \tag{3.664}$$

$$\frac{SEA}{SEA_0} = 5. \tag{3.665}$$

In order for the 'optimized' beam to have the same cross section at every point x, there should be a constant bending moment distribution. This could be achieved by a single moment at the free end.

2.3.33 **Optimization of a simply supported beam under constant distributed load: box cross section**

• Moment distribution:

The bending moment distribution can be taken from the solution of Exercise 2.2.12:

$$M_y(x) = \frac{q_0 L^2}{2}\left(\left[\frac{x}{L}\right]^2 - \left[\frac{x}{L}\right]^1\right) < 0. \tag{3.666}$$

The maximum bending moment is obtained in the middle of the beam as:

$$M_{y,\max} = M_y\left(x = \frac{L}{2}\right) = -\frac{q_0 L^2}{8}. \tag{3.667}$$

• Geometric properties of the cross section (I_y, A):

For a box profile with external dimensions $a/2 \times a$ and constant wall thickness $d = a/6$ the following results:

$$I_y = \frac{1}{12}\frac{a}{2}a^3 - \frac{1}{12}\left(\frac{a}{2} - 2\frac{a}{6}\right)\left(a - 2\frac{a}{6}\right)^3 \tag{3.668}$$

$$= \frac{a^3}{24} - \frac{1}{12}\frac{a}{6}\left(\frac{2a}{3}\right)^3 = \frac{73}{1944}a^4, \tag{3.669}$$

$$A = a\frac{a}{2} - \left(\frac{a}{2} - 2\frac{a}{6}\right)\left(a - 2\frac{a}{6}\right) \tag{3.670}$$

$$= \frac{7}{18}a^2. \tag{3.671}$$

- Beam contour $a = a(x)$ along the beam axis:

Using the stress criterion

$$\sigma_{x,\mathrm{max}} = \frac{M_y(x)}{I_y(x)} \times \frac{a(x)}{2} = \frac{\frac{q_0 L^2}{2}\left(\left[\frac{x}{L}\right] - \left[\frac{x}{L}\right]^2\right)}{\frac{73 a(x)^4}{1944}} \times \frac{a(x)}{2} \stackrel{!}{=} R_{\mathrm{p0.2}}, \tag{3.672}$$

results

$$a^3(x) = \frac{1944 q_0 L^2}{4 \times 73 R_{\mathrm{p0.2}}} \times \left(\left[\frac{x}{L}\right] - \left[\frac{x}{L}\right]^2\right). \tag{3.673}$$

The evaluation of the stress criterion in the reference cross section at the location $x = L/2$ provides the reference dimension to:

$$a_0^3 = \frac{1944 q_0 L^2}{4 \times 4 \times 73 R_{\mathrm{p0.2}}}. \tag{3.674}$$

With this result we get (see Fig. 3.33):

$$a(x) = (4)^{\frac{1}{3}}\, a_0 \left(-\left(\frac{x}{L}\right)^2 + \frac{x}{L}\right)^{\frac{1}{3}} \approx 1.587 \times a_0 \left(-\left(\frac{x}{L}\right)^2 + \frac{x}{L}\right)^{\frac{1}{3}} > 0, \tag{3.675}$$

where the following dimensions result for the reference cross section:

$$a_0 = \left(\frac{486 q_0 L^2}{4 \times 73 R_{\mathrm{p0.2}}}\right)^{\frac{1}{3}}. \tag{3.676}$$

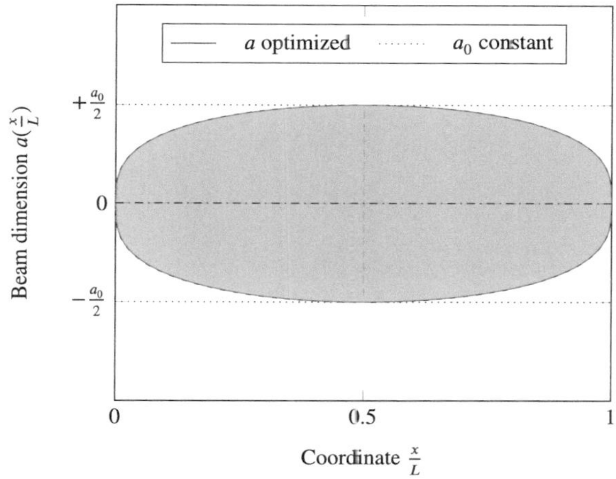

Fig. 3.33 Beam dimension a along the beam axis

2.3.34 **Optimization of a simply supported bending under constant distributed load: box cross section. Numerical calculation of the specific energy absorption**

• Moment distribution:

The bending moment distribution can be taken from the solution of Exercise 2.3.33:

$$M_y(x) = \frac{q_0 L^2}{2}\left(\left[\frac{x}{L}\right]^2 - \left[\frac{x}{L}\right]^1\right) < 0. \tag{3.677}$$

The maximum bending moment is obtained in the middle of the beam as:

$$M_{y,\text{max}} = M_y\left(x = \frac{L}{2}\right) = -\frac{q_0 L^2}{8}. \tag{3.678}$$

• Geometric properties of the cross section (I_y, A):

For a box profile with external dimensions $a/2 \times a$ and constant wall thickness $d = a/6$ we get (see Exercise 2.3.33):

$$I_y = \frac{73}{1944}a^4, \tag{3.679}$$

$$A = \frac{7}{18}a^2. \tag{3.680}$$

• Constant cross section:

The total strain energy is calculated according to Eq. (3.32)$_{TB}$ for a bending load case as follows:

$$\Pi = \int_0^L \frac{M_y(x)^2}{2EI_y} \mathrm{d}x. \tag{3.681}$$

For our example, the total strain energy for the reference cross section is obtained by integration as:

$$\Pi_0 = \frac{1}{2E} \int_0^L \frac{\frac{q_0^2 L^4}{4}\left(\left[\frac{x}{L}\right] - \left[\frac{x}{L}\right]^2\right)^2}{\frac{73}{1944} a_0^4} \mathrm{d}x, \tag{3.682}$$

$$= \frac{1944 q_0^2 L^4}{8 \times 73 E a_0^4} \underbrace{\int_0^L \left(\left[\tfrac{x}{L}\right] - \left[\tfrac{x}{L}\right]^2\right)^2 \mathrm{d}x}_{\frac{L}{30}}, \tag{3.683}$$

$$= \frac{81 q_0^2 L^5}{730 E a_0^4}. \tag{3.684}$$

The specific energy absorption is generally given by Eq. (3.27)$_{TB}$ as:

$$SEA = \frac{\Pi}{m}, \tag{3.685}$$

or taking into account the expressions for the total strain energy (see Eq. (3.684)) and mass ($m = \varrho L \frac{7}{18} a_0^2$):

$$SEA_0 = \frac{729 q_0^2 L^4}{2555 E a_0 \varrho} \approx 0.285323 \times \frac{q_0^2 L^4}{E a_0 \varrho}. \tag{3.686}$$

In summary, one obtains for the reference beam:

$$A_0 = \frac{7 a_0{}^2}{18}, \tag{3.687}$$

$$I_0 = \frac{73 a_0{}^4}{1944}, \tag{3.688}$$

$$m_0 = \frac{8\, L\, a_0{}^2 \varrho}{9}, \tag{3.689}$$

$$\Pi_0 = \frac{81 L^5\, q_0{}^2}{730 E\, a_0{}^4}, \tag{3.690}$$

$$SEA_0 = \frac{729 L^4\, q_0{}^2}{2555 E\, a_0{}^6 \varrho}. \tag{3.691}$$

- Optimized cross section:

The total strain energy for the optimized beam is obtained by integration as:

$$\Pi = \frac{1}{2E}\int_0^L \frac{\frac{q_0^2 L^4}{4}\left(\left[\frac{x}{L}\right]-\left[\frac{x}{L}\right]^2\right)^2}{\frac{73}{1944}4^{4/3}a_0^4\left(\left[\frac{x}{L}\right]-\left[\frac{x}{L}\right]^2\right)^{\frac{4}{3}}}\mathrm{d}x \tag{3.692}$$

$$= \frac{1944}{8\times 73\times 4^{4/3}}\times\frac{q_0^2L^4}{Ea_0^4}\underbrace{\int_0^L\left(\left[\frac{x}{L}\right]-\left[\frac{x}{L}\right]^2\right)^{\frac{2}{3}}\mathrm{d}x}_{0.2933415685302212\times L\ (n=4112)} \tag{3.693}$$

$$0.1537837221422918\times\frac{q_0^2\,L^5}{Ea_0^4}. \tag{3.694}$$

The volume is determined by evaluating the integral:

$$V=\int_0^L A(x)\,\mathrm{d}x=\int_0^L\frac{7}{18}a^2(x)\,\mathrm{d}x=\frac{7}{18}4^{\frac{2}{3}}\,a_0^2\underbrace{\int_0^L\left(\left[\frac{x}{L}\right]-\left[\frac{x}{L}\right]^2\right)^{\frac{2}{3}}}_{0.2933415685302212\times L\ (n=4112)} \tag{3.695}$$

$$\approx 0.2874567243336484\times a_0^2L. \tag{3.696}$$

In summary, one obtains for the optimized cross section (see Fig. 3.33):

$$a(x)=(4)^{\frac{1}{3}}\,a_0\left(-\left(\frac{x}{L}\right)^2+\frac{x}{L}\right)^{\frac{1}{3}}, \tag{3.697}$$

$$A(x)=\frac{7}{18}a(x)^2, \tag{3.698}$$

$$I(x)=\frac{73}{1944}a(x)^4, \tag{3.699}$$

$$V=\frac{7}{18}\times 4^{\frac{2}{3}}\times a_0^2\times 0.2933415685302212\,L, \tag{3.700}$$

$$m=\frac{7}{18}\times 4^{\frac{2}{3}}\times a_0^2\times\varrho\times 0.2933415685302212\,L, \tag{3.701}$$

$$\Pi=\frac{0.1537837221422918L^5\,{q_0}^2}{E\,{a_0}^4}, \tag{3.702}$$

$$SEA=\frac{0.5349804305283756L^4\,{q_0}^2}{E\,{a_0}^6\varrho}. \tag{3.703}$$

- Ratios:

$$\frac{m}{m_0} = 0.739, \tag{3.704}$$

$$\frac{SEA}{SEA_0} = 1.875. \tag{3.705}$$

2.3.35 **Optimal design of a cantilever beam (classical optimization problem)**

- Geometric properties of the cross section (I_y, A):

For a rectangular cross-section with external dimensions $a \times 2a$ we get:

$$I_y = \frac{1}{12}a(2a)^3 = \frac{2}{3}a^4, \tag{3.706}$$

$$A = 2a^2. \tag{3.707}$$

- Objective function:

The objective function of the optimization problem, i.e. the mass of the beam, can be represented as a function of the design variable $X = a$ as follows:

$$F(X) = m(X) = \varrho V(X) = \varrho L A(X) = 2\varrho L X^2. \tag{3.708}$$

- Inequality constraints:

The minimization is to be carried out taking into account the following inequality constraints, see Öchsner (2014):

The maximum deflection can be taken from the solution to Exercise 2.1.2a. It should be noted that the actual force acts in the positive z-direction:

$$u_{z,\max} = \frac{F_0 L^3}{3EI_y}, \tag{3.709}$$

or formulated as an inequality constraint:

$$g_1(X) = \frac{F_0 L^3}{2EX^4} - \underbrace{u_{z,\max}}_{r_1 L} \leq 0. \tag{3.710}$$

The maximum bending moment can also be taken from the solution to Exercise 2.1.2a. Taking into account the direction of the force F_0, the result is:

$$M_y(x) = -F_0 L \left(1 - \frac{x}{L}\right). \tag{3.711}$$

This gives the maximum stress at the clamping location as:

$$\sigma_{x,\max} = \frac{M_y(0)}{I_y} \times (-a) = \frac{-F_0 L}{\frac{2a^4}{3}} \times (-a) \stackrel{!}{=} R_{p0.2}, \tag{3.712}$$

or formulated as an inequality constraint (with $a = X$):

$$g_2(X) = \frac{3F_0L}{2X^3} - R_{p0.2} \leq 0. \tag{3.713}$$

In summary, the optimization problem is:

Minimize the objective function

$$F(X) = 2\varrho L X^2, \tag{3.714}$$

taking into account the inequality constraints (see Fig. 3.34):

$$g_1(X) = \frac{F_0L^3}{2EX^4} - r_1 L \leq 0 \quad \text{(displacement)}, \tag{3.715}$$

$$g_2(X) = \frac{3F_0L}{2X^3} - R_{p0.2} \leq 0 \quad \text{(stress)}. \tag{3.716}$$

From Fig. 3.34, one can see that the inequality constraint g_1 is the stricter condition. Thus, only points to the right of the blue right-hand boundary curve are allowed on the curve $F(X)$ (this follows from the condition that $g_1 \leq 0$ must be). The optimized design variable results from the condition

$$g_1(X) \stackrel{!}{=} 0, \tag{3.717}$$

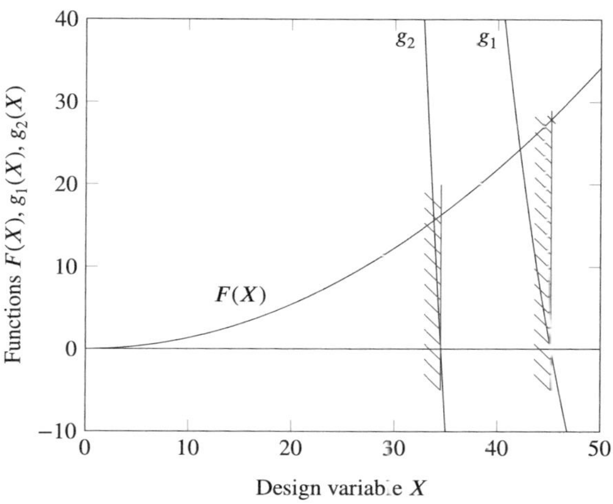

Fig. 3.34 Optimal design of a cantilever beam with: $L = 2540$ mm, $F_0 = 2667$ N, $E = 68948$ MPa, $R_{p0.2} = 247$ MPa, $\varrho = 2.691 \times 10^{-6}$ kg/mm^3, $r_1 = 0.03$ (original set of Eqs. (3.714)–(3.716)). Exact solution for the minimum: $X_{extr} = 45.160$ mm

as analytical solution to:

$$X_{\text{extr}} = \left(\frac{F_0 L^2}{2 E r_1} \right)^{\frac{1}{4}} = 45.16\,\text{mm}. \tag{3.718}$$

As an alternative to Eqs. (3.714)–(3.716), one can also specify a normalized representation, i.e. without a specific unit. By introducing the normalized geometric dimension $a^{\text{n}} = a/L = X^{\text{n}}$, one arrives at the corresponding equations of the optimization problem:

$$F^{\text{n}}(X^{\text{n}}) = F^{\text{n}}(a^{\text{n}}) = \frac{m}{\varrho \times L^3} = 2 \times \left(a^{\text{n}}\right)^2 = 2 \times \left(X^{\text{n}}\right)^2, \tag{3.719}$$

$$g_1^{\text{n}}(X^{\text{n}}) = g_1^{\text{n}}(a^{\text{n}}) = \frac{F_0}{2EL^2 r_1 (a^{\text{n}})^4} - 1 = \frac{F_0}{2EL^2 r_1 (X^{\text{n}})^4} - 1 \leq 0, \tag{3.720}$$

$$g_2^{\text{n}}(X^{\text{n}}) = g_2^{\text{n}}(a^{\text{n}}) = \frac{3F_0}{2L^2 R_{\text{p0.2}} (a^{\text{n}})^3} - 1 = \frac{3F_0}{2L^2 R_{\text{p0.2}} (X^{\text{n}})^3} - 1 \leq 0. \tag{3.721}$$

The graphical representation of these normalized equations is shown in Fig. 3.35a. It can be seen that both inequality constraints again have a single root in the considered region of the abscissa. However, both slopes are extremely steep. Thus, a scaling factor β can be introduced to modify the inequality restrictions as follows:

$$g_1^{\text{n}}(X^{\text{n}})/\beta = \frac{1}{\beta} \left(\frac{F_0}{2EL^2 r_1 (X^{\text{n}})^4} - 1 \leq 0 \right), \tag{3.722}$$

$$g_2^{\text{n}}(X^{\text{n}})/\beta = \frac{1}{\beta} \left(\frac{3F_0}{2L^2 R_{\text{p0.2}} (X^{\text{n}})^3} - 1 \leq 0 \right). \tag{3.723}$$

It is obvious that the scaling does not change the position of the roots and a similar course of all curves is maintained, see Fig. 3.35b.

2.3.36 **Optimal design of a compression strut (classical optimization problem)**

- Geometric properties of the cross section (I_y, A):

For a rectangular cross section with external dimensions $a \times 2a$ we get:

$$I = \frac{1}{12} a (2a)^3 = \frac{2}{3} a^4 \quad \text{oder} \quad \frac{1}{12} 2a (a)^3 = \frac{1}{6} a^4, \tag{3.724}$$

$$A = 2a^2. \tag{3.725}$$

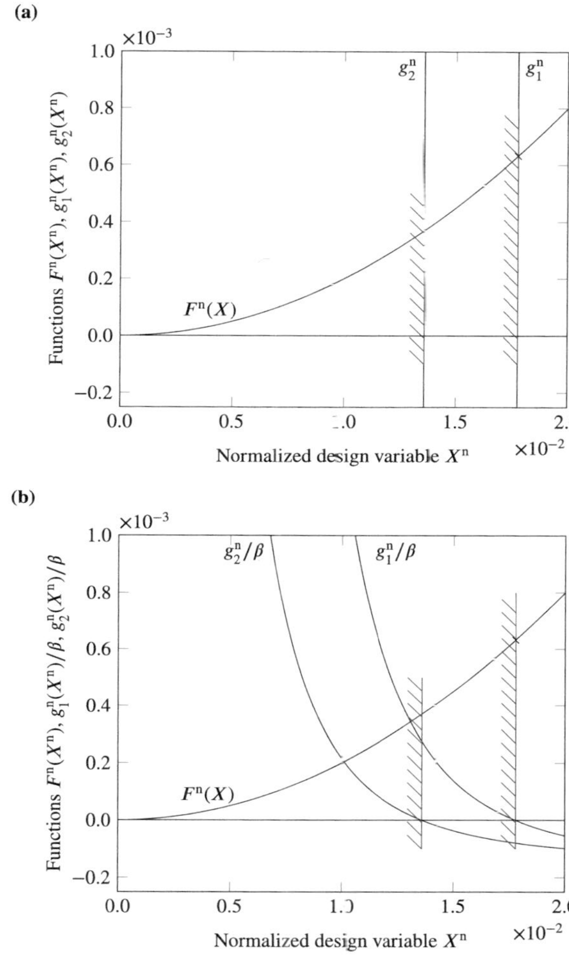

Fig. 3.35 Optimal design of a cantilever beam: **a** normalized equations and **b** normalized equations with scaled inequality constraints ($\beta = 7000$). Exact solution for the minimum: $X^n_{\text{extr}} = 0.01778$

• Objective function:

The objective function of the optimization problem, i.e. the mass of the strut, can be represented as a function of the design variable $X = a$ as follows:

$$F(X) = m(X) = \varrho V(X) = \varrho L A(X) = 2\varrho L X^2. \tag{3.726}$$

• Inequality constraints:

The magnitude of the maximum stress is:

$$\sigma_{x,\max} = \frac{F_0}{A} \overset{!}{=} R_{p0.2}, \tag{3.727}$$

or formulated as an inequality constraint (with $a = X$):

$$g_1(X) = \frac{F_0}{2X^2} - R_{p0.2} \leq 0. \tag{3.728}$$

The critical buckling load for Euler case 1 is (see Eq. (A.23)):

$$F_{cr} = \frac{\pi^2 E I_{min}}{4L^2}, \tag{3.729}$$

or formulated as an inequality constraint (with $a = X$):

$$g_2(X) = F_0 - \frac{\pi^2 E X^4}{24L^2} \leq 0. \tag{3.730}$$

In summary, the optimization problem is:

Minimize the objective function

$$F(X) = m(X) = 2\varrho L X^2, \tag{3.731}$$

taking into account the inequality constraints (see Fig. 3.36):

$$g_1(X) = \frac{F_0}{2X^2} - R_{p0.2} \leq 0 \quad \text{(stress)}, \tag{3.732}$$

$$g_2(X) = F_0 - \frac{\pi^2 E X^4}{24L^2} \leq 0 \quad \text{(buckling)}. \tag{3.733}$$

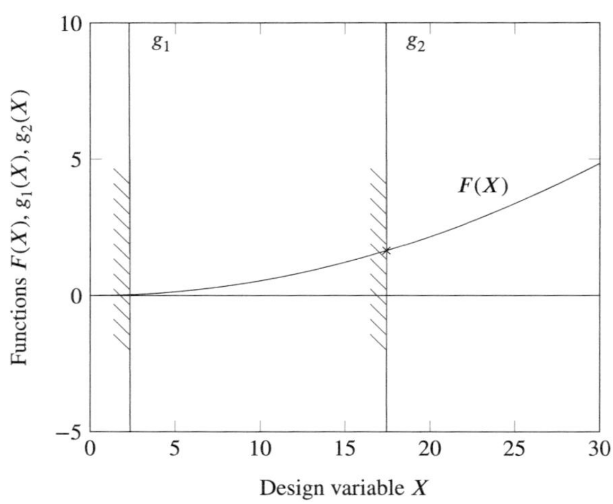

Fig. 3.36 Optimal design of a compression strut: $L = 1000$ mm, $F_0 = 2667$ N, $E = 70000$ MPa, $R_{p0.2} = 247$ MPa, $\varrho = 2.691 \times 10^{-6}$ kg/mm^3 (original set of Eqs. (3.731)–(3.733)). Exact solution for the minimum: $X_{extr} = 17.447$ mm

It can be seen from Fig. 3.36 that the inequality constraint g_2 is the stricter condition. Thus, only points to the right of the blue right-hand boundary curve are allowed on the curve $F(X)$ (this results from the condition that $g_2 \leq 0$ must be). The optimized design variable results from the condition

$$g_2(X) \stackrel{!}{=} 0, \tag{3.734}$$

as analytical solution to:

$$X_{\text{extr}} = \left(\frac{24 F_0 L^2}{\pi^2 E} \right)^{\frac{1}{4}} = 17.45 \text{ mm}. \tag{3.735}$$

2.3.37 **Optimal design of a short cantilever beam (classical optimization problem)**

- Internal reactions (Q_z, M_y):

The functions of the internal reactions can be taken from the solution of Exercise 2.1.2:

$$Q_z(x) = -F_0, \tag{3.736}$$

$$M_y(x) = F_0 L \left(1 - \frac{x}{L} \right), \tag{3.737}$$

where the maximum bending moment is obtained at the clamping point ($x = 0$) as:

$$M_{y,\max} = M_y\,(x = 0) = F_0 L. \tag{3.738}$$

- Geometric properties of the cross section (I_y, A):

For a rectangular cross section with external dimensions $a \times 2a$ we get:

$$I_y = \frac{1}{12} a (2a)^3 = \frac{2}{3} a^4, \tag{3.739}$$

$$A = 2a^2. \tag{3.740}$$

- Objective function:

The objective function of the optimization problem, i.e. the mass of the beam, can be represented as a function of the design variable $X = a$ as follows:

$$F(X) = m(X) = \varrho V(X) = \varrho L A(X) = 2 \varrho L X^2. \tag{3.741}$$

• Inequality constraints:

The minimization is to be carried out taking into account the following inequality constraints, see Öchsner (2014):

The maximum normal stress is obtained at the clamping location as:

$$\sigma_{x,\max} = \frac{M_y(0)}{I_y} \times (-a) = \frac{F_0 L}{\frac{2a^4}{3}} \times a \stackrel{!}{=} R_{\text{p0.2}}, \tag{3.742}$$

or formulated as an inequality constraint (with $a = X$):

$$g_1(X) = \frac{3F_0 L}{2X^3} - R_{\text{p0.2}} \leq 0. \tag{3.743}$$

The maximum shear stress is obtained according to Eq. $(2.17)_{\text{TB}}$ in the middle of the cross section as:

$$\tau_{xz,\max}(x, z) = \tau_{xz}(x, z = 0) = \frac{Q_z(x)}{2I_y} \times (a)^2 = \frac{R_{\text{p0.2}}}{2}, \tag{3.744}$$

or formulated as an inequality constraint (with $a = X$):

$$g_2(X) = \frac{3F_0}{4X^2} - \frac{R_{\text{p0.2}}}{2} \leq 0. \tag{3.745}$$

In summary, the optimization problem is:

Minimize the objective function

$$F(X) = 2\varrho L X^2, \tag{3.746}$$

taking into account the inequality constraints (see Fig. 3.37):

$$g_1(X) = \frac{3F_0 L}{2X^3} - R_{\text{p0.2}} \leq 0 \quad \text{(normal stress)}, \tag{3.747}$$

$$g_2(X) = \frac{3F_0}{4X^2} - \frac{R_{\text{p0.2}}}{2} \leq 0 \quad \text{(shear stress)}. \tag{3.748}$$

It can be seen from Fig. 3.37 that the inequality constraint g_1 is the stricter condition. Thus, only points to the right of the blue right-hand boundary curve are allowed on the curve $F(X)$ (this follows from the condition that $g_1 \leq 0$ must be). The optimized design variable results from the condition

$$g_1(X) \stackrel{!}{=} 0, \tag{3.749}$$

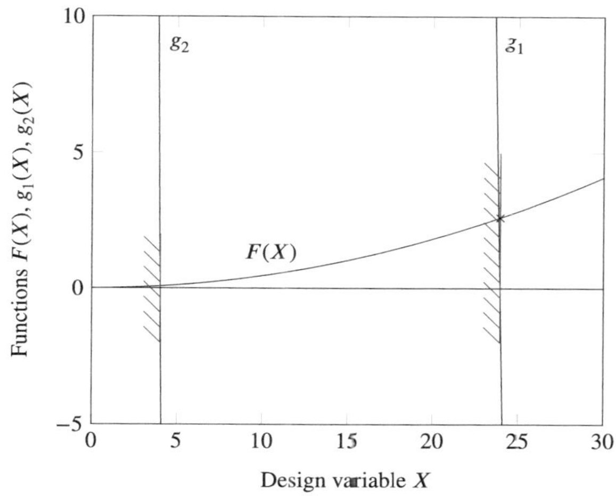

Fig. 3.37 Optimal design of a short cantilever beam: $L = 846.67$ mm, $F_0 = 2667$ N, $E = 68948$ MPa, $R_{p0.2} = 247$ MPa, $\varrho = 2.691 \times 10^{-6}$ kg/mm^3, (original set of Eqs. (3.746)–(3.748)). Exact solution for the minimum: $X_{extr} = 23.936$ mm

as analytical solution to:

$$X_{extr} = \left(\frac{3F_0L}{2R_{p0.2}}\right)^{\frac{1}{3}} = 23.94 \text{ mm}. \tag{3.750}$$

2.3.38 **Optimal design of a cantilever beam: constant rectangular cross section (classical optimization problem)**

- Geometric properties of the cross section (I_y, A):

For a rectangular cross-section with external dimensions $a \times b$ we get:

$$I_y = \frac{1}{12}a(b)^3 = \frac{1}{12}ab^3, \tag{3.751}$$

$$A = ab. \tag{3.752}$$

- Objective function:

The objective function of the optimization problem, i.e. the mass of the beam, can be represented as a function of the two design variables $X_1 = a$ and $X_2 = b$ as follows:

$$F(X_1, X_2) = m(X_1, X_2) = \varrho V(X_1, X_2) = \varrho L X_1 X_2, \tag{3.753}$$

- Inequality constraints:

The minimization must be carried out taking into account the following three inequality constraints, see Öchsner (2014):

The maximum deflection can be taken from the solution to Exercise 2.1.2a. It should be noted that the actual force acts in the positive z-direction:

$$u_{z,\max} = \frac{F_0 L^3}{3EI_y}, \tag{3.754}$$

or formulated as an inequality constraint (with $a = X_1$ and $b = X_2$):

$$g_1(X_1, X_2) = \frac{4F_0L^3}{EX_1X_2^3} - \underbrace{u_{z,\max}}_{r_1 L} \leq 0. \tag{3.755}$$

The maximum bending moment can also be taken from the solution to Exercise 2.1.2a. Taking into account the direction of the force F_0, the result is:

$$M_y(x) = -F_0 L \left(1 - \frac{x}{L}\right). \tag{3.756}$$

This gives the maximum stress at the clamping location as:

$$\sigma_{x,\max} = \frac{M_y(0)}{I_y} \times \left(-\frac{b}{2}\right) = \frac{-F_0L}{\frac{ab^3}{12}} \times \left(-\frac{b}{2}\right) \stackrel{!}{=} R_{\text{p0.2}}, \tag{3.757}$$

or formulated as an inequality constraint (with $a = X_1$ and $b = X_2$):

$$g_2(X_1, X_2) = \frac{6F_0L}{X_1X_2^2} - R_{\text{p0.2}} \leq 0. \tag{3.758}$$

The height-to-width ratio $b \leq 20a$ can be formulated as the following inequality constraint (with $a = X_1$ and $b = X_2$):

$$g_3(X_1, X_2) = X_2 - 20X_1 \leq 0. \tag{3.759}$$

In summary, the optimization problem is:

Minimize the objective function

$$F(X_1, X_2) = \varrho L X_1 X_2, \tag{3.760}$$

taking into account the inequality constraints (see Fig. 3.38):

$$g_1(X_1, X_2) = \frac{4F_0L^3}{EX_1X_2^3} - r_1 L \leq 0 \quad \text{(displacement)}, \tag{3.761}$$

$$g_2(X_1, X_2) = \frac{6F_0L}{X_1X_2^2} - R_{\text{p0.2}} \leq 0 \quad \text{(normal stress)}, \tag{3.762}$$

$$g_3(X_1, X_2) = X_2 - 20X_1 \leq 0 \quad \text{(height-to-width ratio)}. \tag{3.763}$$

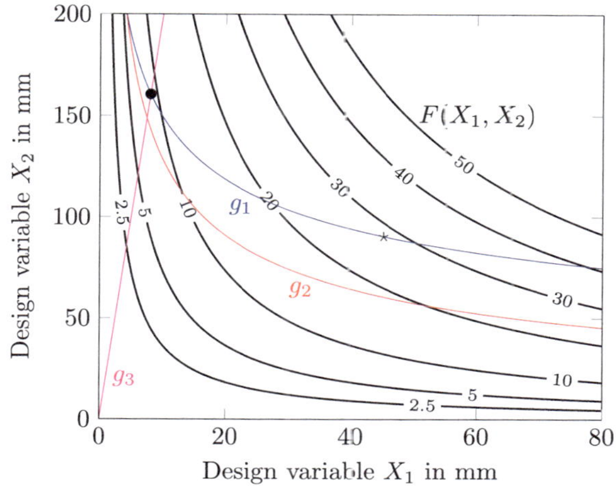

Fig. 3.38 Optimal design of a cantilever beam with: $L = 2540$ mm, $F_0 = 2667$ N, $E = 68948$ MPa, $R_{p0.2} = 247$ MPa, $\varrho = 2.691 \times 10^{-6}$ kg/mm^3, $r_1 = 0.03$ (original set of Eqs. (3.760)–(3.763)). Exact solution for the minimum: $X_{1,\text{extr}} = 8.031$ mm, $X_{2,\text{extr}} = 160.614$ mm (indicated by the • marker)

From Fig. 3.38 one can see that the minimum—taking into account the given inequality constraints—is the intersection point of the curves g_1 and g_3. A simple calculation gives the analytical solution to:

$$X_{1,\text{extr}} = (20)^{-\frac{3}{4}} \times \left(\frac{4F_0L^2}{Er_1}\right)^{\frac{1}{4}} = 8.031 \text{ mm}, \tag{3.764}$$

$$X_{2,\text{extr}} = 20 \times X_{1,\text{extr}} = 160.614 \text{ mm}. \tag{3.765}$$

The result of Problem 2.3.35, i.e. $a = 45.160$ and $2a = 90.3$, is marked by the ⋆ marker in Fig. 3.38. For the given geometric dimensions this results in a total mass of $F(a, 2a) = m(a, 2a) = 27.9$ kg, while the optimization for a variable width and height results in $F(X_{1,\text{extr}}, X_{2,\text{extr}}) = m(X_{1,\text{extr}}, X_{2,\text{extr}}) = 8.8$ kg.

2.3.39 **Optimal design of a simply supported beam: constant rectangular cross section (classical optimization problem)**

- Geometric properties of the cross section (I_y, A):

For a rectangular cross section with external dimensions $a \times b$ we get:

$$I_y = \frac{1}{12}a(b)^3 = \frac{1}{12}ab^3, \tag{3.766}$$

$$A = ab. \tag{3.767}$$

- Internal reactions (M_y, Q_z):

The functions of the internal reactions can be taken from the solution to the Exercise 2.2.13. Note that in the current problem, the force F_C acts in the positive z-direction:

$$M_y(x) = +\frac{F_0}{2} \times x \quad \text{for } 0 \leq x \leq \frac{L}{2}, \tag{3.768}$$

$$Q_z(x) = +\frac{F_0}{2} \quad \text{for } 0 \leq x \leq \frac{L}{2}. \tag{3.769}$$

- Objective function:

The objective function of the optimization problem, i.e. the mass of the beam, can be represented as a function of the two design variables $X_1 = a$ and $X_2 = b$ as follows:

$$F(X_1, X_2) = m(X_1, X_2) = \varrho V(X_1, X_2) = \varrho L X_1 X_2, \tag{3.770}$$

- Inequality constraints:

The minimization must be carried out taking into account the following four inequality constraints, see Öchsner (2014):

The maximum normal stress is obtained in the middle of the beam as:

$$\sigma_{x,\max} = \frac{M_y(L/2)}{I_y} \times \frac{b}{2} = \frac{\frac{F_0 L}{4}}{\frac{ab^3}{12}} \times \frac{b}{2} \stackrel{!}{=} R_{\text{p}0.2}, \tag{3.771}$$

or formulated as an inequality constraint (with $a = X_1$ and $b = X_2$):

$$g_1(X_1, X_2) = \frac{3F_0 L}{2X_1 X_2^2} - R_{\text{p}0.2} \leq 0. \tag{3.772}$$

The maximum shear stress is obtained according to Eq. $(2.17)_{\text{TB}}$ in the middle of the cross section as:

$$\tau_{xz,\max}(x, z) = \tau_{xz}(x, z = 0) = \frac{Q_z(x)}{2I_y} \times \left(\frac{b}{2}\right)^2 = \frac{R_{\text{p}0.2}}{2}, \tag{3.773}$$

or formulated as an inequality constraint (with $a = X_1$ and $b = X_2$):

$$g_2(X_1, X_2) = \frac{3F_0}{4X_1 X_2} - \frac{R_{\text{p}0.2}}{2} \leq 0. \tag{3.774}$$

The maximum deflection in the middle of the beam is obtained by determining the bending line. Using the general solutions according to Eqs. $(2.12)_{\text{TB}}$–$(2.15)_{\text{TB}}$, i.e., for $q_0 = 0$ and constant material and geometric properties

$$Q_z(x) = -c_1, \tag{3.775}$$

$$M_y(x) = -c_1 x - c_2, \tag{3.776}$$

$$\varphi_y(x) = -\frac{1}{EI_y}\left(\frac{c_1 x^2}{2} + c_2 x + c_3\right), \tag{3.777}$$

$$u_z(x) = \frac{1}{EI_y}\left(\frac{c_1 x^3}{6} + \frac{c_2 x^2}{2} + c_3 x + c_4\right), \tag{3.778}$$

the integration constants, taking into account the boundary conditions $u_z(0) = 0$, $M_y(0) = 0$, $Q_z(0) = F_0/2$ and $\varphi_y(L/2) = 0$, are:

$$c_4 = 0, \tag{3.779}$$

$$c_2 = 0, \tag{3.780}$$

$$c_1 = -\frac{F_0}{2}, \tag{3.781}$$

$$c_3 = \frac{F_0 L^2}{16}. \tag{3.782}$$

Thus the bending line is:

$$u_z(x) = \frac{F_0 L^3}{EI_y}\left(-\frac{1}{12}\left[\frac{x}{L}\right]^3 + \frac{1}{16}\left[\frac{x}{L}\right]^1\right), \tag{3.783}$$

respectively the maximum deflection to:

$$u_{z,\max} = u_z(x = L/2) = \frac{F_0 L^3}{48EI_y}. \tag{3.784}$$

From this, the inequality restriction can be written as (with $a = X_1$ and $b = X_2$):

$$g_3(X_1, X_2) = \frac{F_0 L^3}{4EX_1X_2^3} - r_1 L \leq 0. \tag{3.785}$$

The height-to-width ratio $b \leq 20a$ can be formulated as the following inequality constraint (with $a = X_1$ and $b = X_2$):

$$g_3(X_1, X_2) = X_2 - 20X_1 \leq 0. \tag{3.786}$$

In summary, the optimization problem is:

Minimize the objective function

$$F(X_1, X_2) = \varrho L X_1 X_2, \tag{3.787}$$

taking into account the inequality constraints (see Fig. 3.39):

$$g_1(X_1, X_2) = \frac{3F_0L}{2X_1X_2^2} - R_{p0.2} \leq 0 \quad \text{(normal stress)}, \tag{3.788}$$

$$g_2(X_1, X_2) = \frac{3F_0}{4X_1X_2} - \frac{R_{p0.2}}{2} \leq 0 \quad \text{(shear stress)}, \tag{3.789}$$

$$g_3(X_1, X_2) = \frac{F_0L^3}{4EX_1X_2^3} - r_1L \leq 0 \quad \text{(displacement)}, \tag{3.790}$$

$$g_4(X_1, X_2) = X_2 - 20X_1 \leq 0 \quad \text{(height-to-width ratio)}. \tag{3.791}$$

From Fig. 3.39 one can see that the minimum—taking into account the given inequality constraints—is the intersection of the curves g_1 and g_4 for the case $r_1 = 0.03$ and the intersection of the curves g_3 and g_4 for the case $r_1 = 0.003$. A simple calculation gives the analytical solution to:

$$X_{1,\text{extr}}\Big|_{r_1=0.03} = (20)^{-\frac{2}{3}} \times \left(\frac{3F_0L}{2R_{p0,2}}\right)^{\frac{1}{3}} = 4.685 \text{ mm}, \tag{3.792}$$

$$X_{2,\text{extr}}\Big|_{r_1=0.03} = 20 \times X_{1,\text{extr}} = 93.704 \text{ mm}, \tag{3.793}$$

respectively

$$X_{1,\text{extr}}\Big|_{r_1=0.003} = (20)^{-\frac{3}{4}} \times \left(\frac{F_0L^2}{4Er_1}\right)^{\frac{1}{4}} = 7.140 \text{ mm}, \tag{3.794}$$

$$X_{2,\text{extr}}\Big|_{r_1=0.003} = 20 \times X_{1,\text{extr}} = 142.809 \text{ mm}. \tag{3.795}$$

2.3.40 **Optimal design of a compression strut: constant rectangular cross section (classical optimization problem)**

- Geometric properties of the cross section (I_y, A):

For a rectangular cross section with external dimensions $a \times b$ we get:

$$I_y = \frac{1}{12}a(b)^3, \tag{3.796}$$

$$I_z = \frac{1}{12}b(a)^3, \tag{3.797}$$

$$A = ab. \tag{3.798}$$

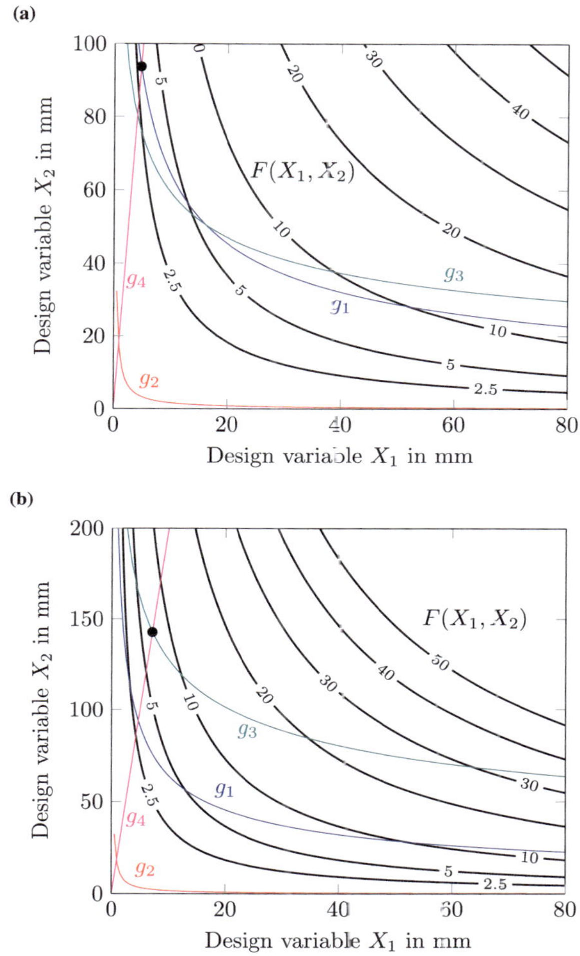

Fig. 3.39 Optimal design of a simply supported beam with $L = 2540\,\text{mm}$, $F_0 = 2667\,\text{N}$, $E = 68948\,\text{MPa}$, $R_{\text{p0.2}} = 247\,\text{MPa}$, $\varrho = 2.691 \times 10^{-6}\,\text{kg/mm}^3$ (original set of Eqs. (3.787)–(3.791)): **a** $r_1 = 0.03$, exact solution for the minimum: $X_{1,\text{extr}} = 4.685$ mm, $X_{2,\text{extr}} = 93.704$ mm (indicated by the • marker), **b** $r_1 = 0.003$, exact solution for the minimum: $X_{1,\text{extr}} = 7.140$ mm, $X_{2,\text{extr}} = 142.809$ mm (indicated by the • marker)

- Objective function:

The objective function of the optimization problem, i.e. the mass of the compression strut, can be represented as a function of the two design variables $X_1 = a$ and $X_2 = b$ as follows:

$$F(X_1, X_2) = m(X_1, X_2) = \varrho V(X_1, X_2) = \varrho L X_1 X_2, \tag{3.799}$$

- Inequality constraints:

The minimization must be carried out taking into account the following three inequality constraints, see Öchsner (2014):

The magnitude of the maximum stress is:

$$\sigma_{x,\max} = \frac{F_0}{A} \overset{!}{=} R_{\text{p0.2}}, \tag{3.800}$$

or formulated as an inequality constraint (with $a = X_1$ and $b = X_2$):

$$g_1(X_1, X_2) = \frac{F_0}{X_1 X_2} - R_{\text{p0.2}} \leq 0. \tag{3.801}$$

The critical buckling load for Euler case 1 for buckling around the y-axis is (see Eq. (A.23)):

$$F_{\text{cr}} = \frac{\pi^2 E I_y}{4L^2}, \tag{3.802}$$

or formulated as an inequality constraint (with $a = X_1$ and $b = X_2$):

$$g_2(X_1, X_2) = F_0 - \frac{\pi^2 E X_1 X_2^3}{48L^2} \leq 0. \tag{3.803}$$

The critical buckling load for Euler case 1 for buckling around the z-axis is (see Eq. (A.23)):

$$F_{\text{cr}} = \frac{\pi^2 E I_z}{4L^2}, \tag{3.804}$$

or formulated as an inequality constraint (with $a = X_1$ and $b = X_2$):

$$g_3(X_1, X_2) = F_0 - \frac{\pi^2 E X_2 X_1^3}{48L^2} \leq 0. \tag{3.805}$$

In summary, the optimization problem is:

Minimize the objective function

$$F(X_1, X_2) = m(X_1, X_2) = \varrho V(X_1, X_2) = \varrho L X_1 X_2, \tag{3.806}$$

taking into account the inequality constraints (see Fig. 3.40):

$$g_1(X_1, X_2) = \frac{F_0}{X_1 X_2} - R_{\text{p0.2}} \leq 0 \quad \text{(normal stress)}, \tag{3.807}$$

$$g_2(X_1, X_2) = F_0 - \frac{\pi^2 E X_1 X_2^3}{48L^2} \leq 0 \quad \text{(buckling around the } y\text{-axis)}, \tag{3.808}$$

$$g_3(X_1, X_2) = F_0 - \frac{\pi^2 E X_2 X_1^3}{48L^2} \leq 0 \quad \text{(buckling around the } z\text{-axis)}. \tag{3.809}$$

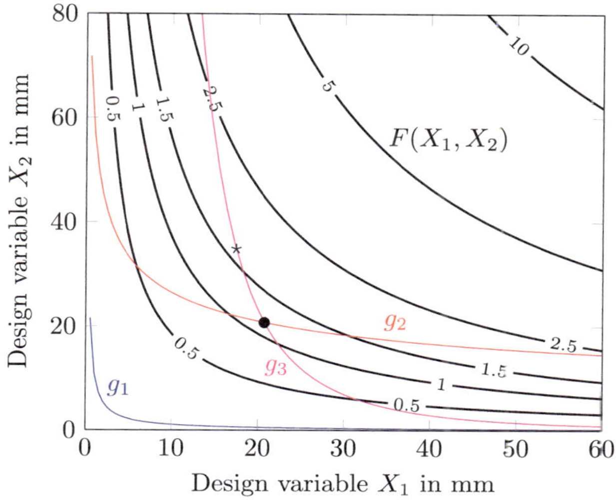

Fig. 3.40 Optimal design of a compression strut: $L = 1000\,\text{mm}$, $F_0 = 2667\,\text{N}$, $E = 70000\,\text{MPa}$, $R_{p0.2} = 247\,\text{MPa}$, $\varrho = 2.691 \times 10^{-6}\,\text{kg/mm}^3$, (original set of Eqs. (3.806)–(3.809)). Exact solution for the minimum: $X_{1,\text{extr}} = 20.748$ mm, $X_{2,\text{extr}} = 20.748$ mm (indicated by the • marker)

From Fig. 3.40 one can see that the minimum—taking into account the given inequality constraints—is the intersection point of the curves g_2 and g_3. A simple calculation gives the analytical solution:

$$X_{1,\text{extr}} = \left(\frac{48 F_0 L^2}{\pi^2 E}\right)^{\frac{1}{4}} = 20.748 \text{ mm}, \tag{3.810}$$

$$X_{2,\text{extr}} = \frac{48 F_0 L^2}{\pi^2 E X_{1,\text{extr}}^3} = 20.748 \text{ mm}. \tag{3.811}$$

The result of Problem 2.3.36, i.e. $a = 17.447$ and $2a = 34.894$, is marked by the ⋆ marker in Fig. 3.40. For the given geometric dimensions this results in a total mass of $F(a, 2a) = m(a, 2a) = 1.638$ kg, while the optimization for a variable width and height results in $F(X_{1,\text{extr}}, X_{2,\text{extr}}) = m(X_{1,\text{extr}}, X_{2,\text{extr}}) = 1.158$ kg.

2.3.41 **Optimal design of a short cantilever beam: constant rectangular cross section (classical optimization problem)**

• Internal reactions (M_y, Q_z):

The functions of the internal reactions can be found in the solution of Exercise 2.1.2:

$$Q_z(x) = -F_0, \tag{3.812}$$

$$M_y(x) = F_0 L \left(1 - \frac{x}{L}\right), \tag{3.813}$$

where the maximum bending moment is obtained at the clamping location ($x = 0$) as:

$$M_{y,\max} = M_y\,(x = 0) = F_0 L. \tag{3.814}$$

- Geometric properties of the cross section (I_y, A):

For a rectangular cross section with external dimensions $a \times b$ we get:

$$I_y = \frac{1}{12} a (b)^3, \tag{3.815}$$

$$A = ab. \tag{3.816}$$

- Objective function:

The objective function of the optimization problem, i.e. the mass of the beam, can be represented as a function of the two design variables $X_1 = a$ and $X_2 = b$ as follows:

$$F(X_1, X_2) = m(X_1, X_2) = \varrho V(X_1, X_2) = \varrho L X_1 X_2. \tag{3.817}$$

- Inequality constraints:

The minimization must be carried out taking into account the following three inequality constraints, see Öchsner (2014):

The maximum normal stress is obtained at the clamping location as:

$$\sigma_{x,\max} = \frac{M_y(0)}{I_y} \times \frac{b}{2} = \frac{F_0 L}{\frac{ab^3}{12}} \times \frac{b}{2} \overset{!}{=} R_{\text{p}0.2}, \tag{3.818}$$

or formulated as an inequality constraint (with $a = X_1$ and $b = X_2$):

$$g_1(X_1, X_2) = \frac{6 F_0 L}{X_1 X_2^2} - R_{\text{p}0.2} \leq 0. \tag{3.819}$$

The maximum shear stress (magnitude) is obtained according to Eq. $(2.17)_{\text{TB}}$ in the middle of the cross section as:

$$\tau_{xz,\max}(x, z) = \tau_{xz}(x, z = 0) = \frac{Q_z(x)}{2 I_y} \times \left(\frac{b}{2}\right)^2 = \frac{R_{\text{p}0.2}}{2}, \tag{3.820}$$

or formulated as an inequality constraint (with $a = X_1$ and $b = X_2$):

$$g_2(X_1, X_2) = \frac{3 F_0}{2 X_1 X_2} - \frac{R_{\text{p}0.2}}{2} \leq 0. \tag{3.821}$$

The height-to-width ratio $b \leq 20a$ can be formulated as the following inequality constraint (with $a = X_1$ and $b = X_2$):

$$g_3(X_1, X_2) = X_2 - 20X_1 \leq 0. \tag{3.822}$$

In summary, the optimization problem is:
Minimize the objective function

$$F(X_1, X_2) = \varrho L X_1 X_2. \tag{3.823}$$

taking into account the inequality restrictions (see Fig. 3.41):

$$g_1(X_1, X_2) = \frac{6F_0 L}{X_1 X_2^2} - R_{\text{p0.2}} \leq 0 \quad \text{(normal stress)}, \tag{3.824}$$

$$g_2(X_1, X_2) = \frac{3F_0}{2X_1 X_2} - \frac{R_{\text{p0.2}}}{2} \leq 0 \quad \text{(shear stress)}, \tag{3.825}$$

$$g_3(X_1, X_2) = X_2 - 20X_1 \leq 0 \quad \text{(height-to-width ratio)}. \tag{3.826}$$

From Fig. 3.41 one can see that the minimum—taking into account the given inequality constraints—is the intersection point of the curves g_1 and g_3. A simple calculation gives the analytical solution:

$$X_{1,\text{extr}} = 20^{-\frac{2}{3}} \times \left(\frac{6F_0 L}{R_{\text{p0.2}}}\right)^{\frac{1}{3}} = 5.157 \text{ mm}, \tag{3.827}$$

$$X_{2,\text{extr}} = 20 \times X_{1,\text{extr}} = 103.135 \text{ mm}. \tag{3.828}$$

The result of Problem 2.3.37, i.e. $a = 23.936$ and $2a = 47.872$, is marked by the ⋆ marker in Fig. 3.41. For the given geometric dimensions this results in a total mass of

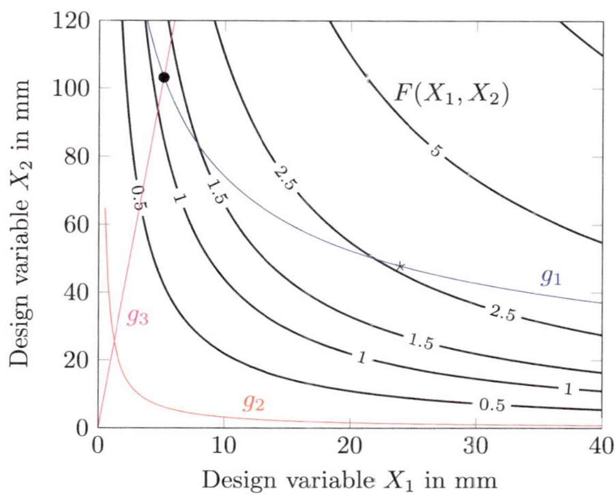

Fig. 3.41 Optimal design of a short cantilever beam: $L = 846.67$ mm, $F_0 = 2667$ N, $E = 68948$ MPa, $R_{\text{p0.2}} = 247$ MPa, $\varrho = 2.691 \times 10^{-6}$ kg/mm^3, (original set of Eqs. (3.823)–(3.826)). Exact solution for the minimum: $X_{1,\text{extr}} = 5.157$ mm, $X_{2,\text{extr}} = 103.135$ mm (indicated by the • marker)

$F(a, 2a) = m(a, 2a) = 2.611$ kg, while the optimization for a variable width and height results in $F(X_{1,\text{extr}}, X_{2,\text{extr}}) = m(X_{1,\text{extr}}, X_{2,\text{extr}}) = 1.212$ kg.

2.3.42 **Optimal design of a stepped cantilever beam with two sections (classical optimization problem)**

The problem requires the calculation of both the deformations and the stress distributions. This task can be approached by a 'hand calculation' based on the finite element method. A single stiffness matrix of an Euler-Bernoulli beam element can be given as follows (see Öchsner (2016, 2018)):

$$\boldsymbol{K}_i^{\text{e}} = \frac{E_i I_i}{L_i^3} \begin{bmatrix} 12 & -6L_i & -12 & -6L_i \\ -6L_i & 4L_i^2 & 6L_i & 2L_i^2 \\ -12 & 6L_i & 12 & 6L_i \\ -6L_i & 2L_i^2 & 6L_i & 4L_i^2 \end{bmatrix}. \tag{3.829}$$

The assembly of the two element stiffness matrices $\boldsymbol{K}_i^{\text{e}}$, taking into account $E_{\text{I}} = E_{\text{II}} = E$, $L_{\text{I}} = L_{\text{II}} = \frac{L}{2}$, $I_{\text{I}} = \frac{1}{12} a b_{\text{I}}^3$ and $I_{\text{II}} = \frac{1}{12} a b_{\text{II}}^3$, results in the following global (reduced) system of equations:

$$E \begin{bmatrix} \left(\frac{8a\,b_{\text{II}}^3}{L^3} + \frac{8a\,b_{\text{I}}^3}{L^3}\right) & \left(\frac{2a\,b_{\text{I}}^3}{L^2} - \frac{2a\,b_{\text{II}}^3}{L^2}\right) & -\frac{8a\,b_{\text{II}}^3}{L^3} & -\frac{2a\,b_{\text{II}}^3}{L^2} \\ \left(\frac{2a\,b_{\text{I}}^3}{L^2} - \frac{2a\,b_{\text{II}}^3}{L^2}\right) & \left(\frac{2a\,b_{\text{II}}^3}{3\,L} + \frac{2a\,b_{\text{I}}^3}{3\,L}\right) & \frac{2a\,b_{\text{II}}^3}{L^2} & \frac{a\,b_{\text{II}}^3}{3\,L} \\ -\frac{8a\,b_{\text{II}}^3}{L^3} & \frac{2a\,b_{\text{II}}^3}{L^2} & \frac{8a\,b_{\text{II}}^3}{L^3} & \frac{2a\,b_{\text{II}}^3}{L^2} \\ -\frac{2a\,b_{\text{II}}^3}{L^2} & \frac{a\,b_{\text{II}}^3}{3\,L} & \frac{2a\,b_{\text{II}}^3}{L^2} & \frac{2a\,b_{\text{II}}^3}{3\,L} \end{bmatrix} \begin{bmatrix} u_{2z} \\ \varphi_{2y} \\ u_{3z} \\ \varphi_{3y} \end{bmatrix} = \begin{bmatrix} 0 \\ 0 \\ -F_0 \\ 0 \end{bmatrix}, \tag{3.830}$$

or in matrix notation as:

$$\boldsymbol{K}\boldsymbol{u} = \boldsymbol{f}. \tag{3.831}$$

Equation (3.830) already takes into account that the two degrees of freedom (i.e. u_{1z} and φ_{1y}) at node 1 are zero. The solution of the system of equations can be obtained, for example, by inverting the global stiffness matrix and then multiplying with the right-hand side, i.e. $\boldsymbol{u} = \boldsymbol{K}^{-1}\boldsymbol{f}$. This gives the column matrix of the unknown node values as:

$$\begin{bmatrix} u_{2z} \\ \varphi_{2y} \\ u_{3z} \\ \varphi_{3y} \end{bmatrix} = \begin{bmatrix} -\frac{5F_0L^3}{4Ea\,b_{\text{I}}^3} \\ \frac{9F_0L^2}{2Ea\,b_{\text{I}}^3} \\ -\frac{F_0L^3\,(7b_{\text{II}}^3+b_{\text{I}}^3)}{2Ea\,b_{\text{I}}^3 b_{\text{II}}^3} \\ \frac{3F_0L^2\,(3b_{\text{II}}^3+b_{\text{I}}^3)}{2Ea\,b_{\text{I}}^3 b_{\text{II}}^3} \end{bmatrix}. \tag{3.832}$$

The bending moment and shear force distributions within a single element of length L_i can generally be expressed using (the index '1' refers to the start node while the index ‘2’ refers to the end node of an element) the following relations:

$$M_y^e(x) = EI_y \left(\left[+\frac{6}{L_i^2} - \frac{12x}{L_i^3} \right] u_{1z} + \left[-\frac{4}{L_i} + \frac{6x}{L_i^2} \right] \varphi_{1y} \right. \\ \left. + \left[-\frac{6}{L_i^2} + \frac{12x}{L_i^3} \right] u_{2z} + \left[-\frac{2}{L_i} + \frac{6x}{L_i^2} \right] \varphi_{2y} \right), \tag{3.833}$$

$$Q_z^e(x) = EI_y \left(\left[-\frac{12}{L_i^3} \right] u_{1z} + \left[+\frac{6}{L_i^2} \right] \varphi_{1y} + \left[+\frac{12}{L_i^3} \right] u_{2z} + \left[+\frac{6}{L_i^2} \right] \varphi_{2y} \right). \tag{3.834}$$

Thus, the normal and shear stress distributions can generally be calculated as follows:

$$\sigma_x^e(x,z) = \frac{M_y^e(x)}{I_y} \times z, \tag{3.835}$$

$$\tau_{xz}^e(x,z) = \frac{Q_z^e(x)}{2I_y} \left[\left(\frac{b}{2} \right)^2 - z^2 \right]. \tag{3.836}$$

The nodal values of the internal reactions for calculating the normal and shear stresses are summarized in Table 3.4. It is noted here that in this case the nodal values based on the finite element calculation are identical to the analytical solution.

From Table 3.4 it can be concluded that the maximum normal stress in each element is reached at the left node and that the shear stress is constant in each element. Thus, the critical stresses in each element can be calculated as follows:

$$\sigma_{x,\mathrm{I}} = \frac{6F_0L}{a\,b_\mathrm{I}^2}, \tag{3.837}$$

$$\sigma_{x,\mathrm{II}} = \frac{3F_0L}{a\,b_\mathrm{II}^2}, \tag{3.838}$$

Table 3.4 Nodal values of the internal reactions bending moment (M_y) and shear force (Q_z) at each node

	Element I		Element II	
	Left node	Right node	Left node	Right node
M_y	F_0L	$\frac{F_0L}{2}$	$\frac{F_0L}{2}$	0
Q_z	$-F_0$	$-F_0$	$-F_0$	$-F_0$

or for the shear stresses:

$$\tau_{xz,\text{I}} = -\frac{3F_0}{2a\,b_\text{I}}, \tag{3.839}$$

$$\tau_{xz,\text{II}} = -\frac{3F_0}{2a\,b_\text{II}}. \tag{3.840}$$

The objective function of the optimization problem, i.e. the mass of the stepped beam, can be represented as a function of the two design variables $X_1 = b_\text{I}$ and $X_2 = b_\text{II}$ as follows:

$$F\,(X_1, X_2) = \varrho L_\text{I} a X_1 + \varrho L_\text{II} a X_2 = \frac{\varrho L a}{2}(X_1 + X_2)\,, \tag{3.841}$$

where the minimization is to be carried out under consideration of the following seven inequality constraints:

$$g_1(X_1, X_2) = +\frac{F_0 L^3 (X_1^3 + 7X_2^3)}{2EaX_1^3X_2^3 r_1 L} - 1 \leq 0 \quad \text{(max. deflection)}, \tag{3.842}$$

$$g_2(X_1, X_2) = \frac{6F_0 L}{aX_1^2} - R_{\text{p}0.2} \leq 0 \quad \text{(normal stress in I)}, \tag{3.843}$$

$$g_3(X_1, X_2) = \frac{3F_0 L}{aX_2^2} - R_{\text{p}0.2} \leq 0 \quad \text{(normal stress in II)}, \tag{3.844}$$

$$g_4(X_1, X_2) = \frac{3F_0}{2aX_1} - \frac{R_{\text{p}0.2}}{2} \leq 0 \quad \text{(shear stress in I)}, \tag{3.845}$$

$$g_5(X_1, X_2) = \frac{3F_0}{2aX_2} - \frac{R_{\text{p}0.2}}{2} \leq 0 \quad \text{(shear stress in II)}, \tag{3.846}$$

$$g_6(X_1, X_2) = X_1 - 20a \leq 0 \quad \text{(height-to-width ratio in I)}, \tag{3.847}$$

$$g_7(X_1, X_2) = X_2 - 20a \leq 0 \quad \text{(height-to-width ratio in II)}. \tag{3.848}$$

The graphical representation of the objective function and the inequality constraints in the X_1-X_2 design space is shown in Figs. 3.42 and 3.43.

From Figs. 3.42 and 3.43 it can be concluded that considering g_1, g_2, and g_3 can be sufficient if a reasonable range of the design variables is chosen. Furthermore, from Fig. 3.42 it can be concluded that the minimum is achieved if g_1 is tangent to the objective function F. In order to derive an analytical solution for the minimum, one must take into account that the objective function F, in the representation $X_2(X_1) = \frac{2c}{\varrho La} - X_1$, represents straight lines with the slope of –1. Thus, the condition for the minimum is that the curve g_1 assumes a slope of –1. Based on Eq. (3.843), the expression for g_1 can be rewritten:

$$X_2 = \left(-\frac{F_0 L^3 X_1^3}{7F_0 L^3 - 2Ear_1 L X_1^3}\right)^{\frac{1}{3}}, \tag{3.849}$$

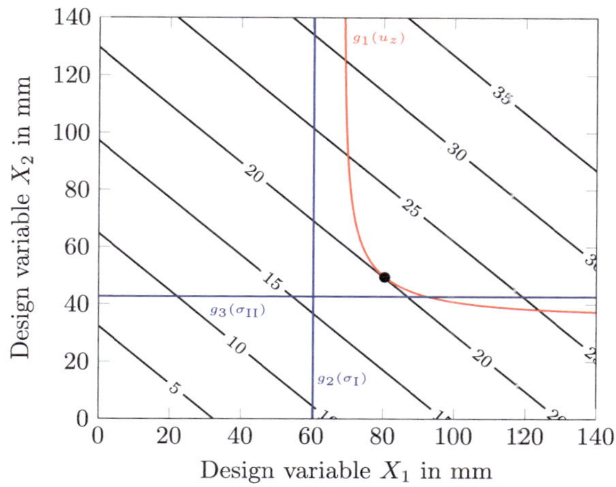

Fig. 3.42 Optimal design of a stepped cantilever beam with two sections: $L = 2540$ mm, $a = 45.16$ mm, $r_1 = 0.06$, $F_0 = 2667$ N, $E = 68948$ MPa, $R_{p0.2} = 247$ MPa, $\varrho = 2.691 \times 10^{-6}$ kg/mm^3, (original set of Eqs. (3.841)–(3.848)). Exact solution for the minimum: $X_{1,\text{extr}} = 80.442$ mm, $X_{2,\text{extr}} = 49.455$ mm (indicated by the • marker)

which, after a short calculation, gives the expression of the first-order derivative as follows:

$$\left.\frac{\mathrm{d}X_2}{\mathrm{d}X_1}\right|_{g_1} = \frac{-7F_0^{\frac{4}{3}}L^{\frac{8}{3}}}{\left(7F_0L^2 - 2Ear_1X_1^3\right)^{\frac{4}{3}}} \stackrel{!}{=} -1. \tag{3.850}$$

The last equation can be rearranged so that one side has a root, and Newton's method (see Sect. A.1.3) can be used to determine the root. It is important that the root is sought in a reasonable interval, e.g. $70 \leq X_{1,\text{ extr}} \leq 85$.

2.3.43 **Optimal design of a simply supported beam with three sections (classical optimization problem)**

The problem requires the calculation of both the deformations and the stress distributions. This task can be approached by a 'hand calculation' based on the finite element method. A single stiffness matrix of an Euler-Bernoulli beam element can be given as follows (see Öchsner (2016, 2018)):

$$\boldsymbol{K}_i^{\mathrm{e}} = \frac{E_i I_i}{L_i^3}\begin{bmatrix} 12 & -6L_i & -12 & -6L_i \\ -6L_i & 4L_i^2 & 6L_i & 2L_i^2 \\ -12 & 6L_i & 12 & 6L_i \\ -6L_i & 2L_i^2 & 6L_i & 4L_i^2 \end{bmatrix}. \tag{3.851}$$

This special problem allows the consideration of symmetry in terms of geometry and loading. Thus, the reduced mechanical system according to Fig. 3.44 allows a faster calculation than when considering the entire structure.

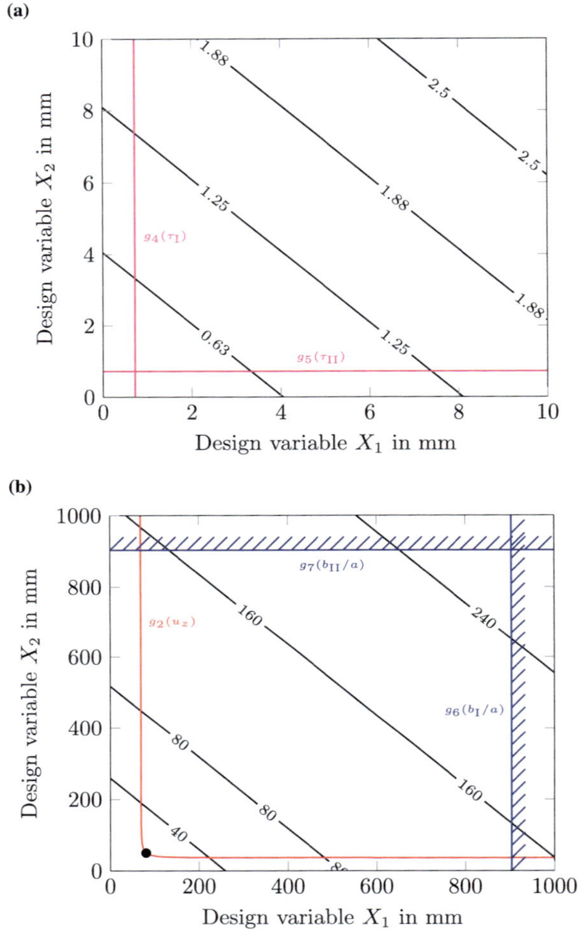

Fig. 3.43 Optimal design of a stepped cantilever beam with two sections: **a** magnification for smaller values of the design variables and **b** global view

The assembly of the two element stiffness matrices $\boldsymbol{K}_i^\text{e}$, taking into account $E_\text{I} = E_\text{II} = E, L_\text{I} = L_\text{II} = \frac{L}{3}, I_\text{I} = \frac{1}{12}ab_\text{I}^3$, and $I_\text{II} = \frac{1}{12}ab_\text{II}^3$, results in the following global (reduced) system of equations:

$$
E\begin{bmatrix}
\frac{a\,b_\text{I}{}^3}{L} & \frac{9a\,b_\text{I}{}^3}{2L^2} & \frac{a\,b_\text{I}{}^3}{2\,L} & 0 \\
\frac{9a\,b_\text{I}{}^3}{2L^2} & \left(\frac{216a\,b_\text{II}{}^3}{L^3}+\frac{27a\,b_\text{I}{}^3}{L^3}\right) & \left(\frac{9a\,b_\text{I}{}^3}{2L^2}-\frac{18a\,b_\text{II}{}^3}{L^2}\right) & -\frac{216a\,b_\text{II}{}^3}{L^3} \\
\frac{a\,b_\text{I}{}^3}{2\,L} & \left(\frac{9a\,b_\text{I}{}^3}{2L^2}-\frac{18a\,b_\text{II}{}^3}{L^2}\right) & \left(\frac{2a\,b_\text{II}{}^3}{L}+\frac{a\,b_\text{I}{}^3}{L}\right) & \frac{18a\,b_\text{II}{}^3}{L^2} \\
0 & -\frac{216a\,b_\text{II}{}^3}{L^3} & \frac{18a\,b_\text{II}{}^3}{L^2} & \frac{216a\,b_\text{II}{}^3}{L^3}
\end{bmatrix}
\begin{bmatrix} \varphi_{1y} \\ u_{2z} \\ \varphi_{2y} \\ u_{3z} \end{bmatrix}
=
\begin{bmatrix} 0 \\ 0 \\ 0 \\ \frac{-F_0}{2} \end{bmatrix}. \tag{3.852}
$$

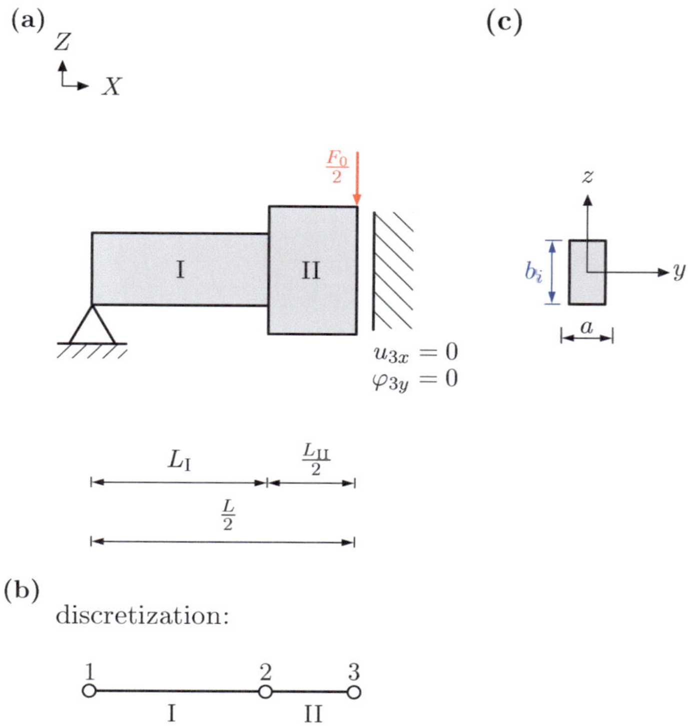

Fig. 3.44 Simply supported beam with three sections under consideration of symmetry: **a** general configuration, **b** discretization, and **c** cross-section of element *i*

The last equation already takes into account that the translational degree of freedom at node 1 and the rotational degree of freedom at node 3 are zero. The solution of the system of equations can be obtained, for example, by inverting the global stiffness matrix and then multiplying with the right-hand side, i.e. $\boldsymbol{u} = \boldsymbol{K}^{-1}\boldsymbol{f}$. This gives the column matrix of the unknown node values as:

$$\begin{bmatrix} \varphi_{1y} \\ u_{2z} \\ \varphi_{2y} \\ u_{3z} \end{bmatrix} = \begin{bmatrix} \frac{F_0 L^2 \left(4{b_\mathrm{II}}^3 + 5{b_\mathrm{I}}^3\right)}{12 E a\, {b_\mathrm{I}}^3 {b_\mathrm{II}}^3} \\ -\frac{F_0 L^3 \left(8{b_\mathrm{II}}^3 + 15{b_\mathrm{I}}^3\right)}{108 E a\, {b_\mathrm{I}}^3 {b_\mathrm{II}}^3} \\ \frac{5 F_0 L^2}{12 E a\, {b_\mathrm{II}}^3} \\ -\frac{F_0 L^3 \left(8{b_\mathrm{II}}^3 + 19{b_\mathrm{I}}^3\right)}{108 E a\, {b_\mathrm{I}}^3 {b_\mathrm{II}}^3} \end{bmatrix}. \tag{3.853}$$

The bending moment and shear force distributions within a single element of length L_i can be generally expressed using (the index '1' refers to the start node, while the index '2' refers to the end node of an element) the following relations:

$$\begin{aligned} M_y^{\mathrm{e}}(x) =& EI_y \left(\left[+\frac{6}{L_i^2} - \frac{12x}{L_i^3} \right] u_{1z} + \left[-\frac{4}{L_i} + \frac{6x}{L_i^2} \right] \varphi_{1y} \right. \\ & \left. + \left[-\frac{6}{L_i^2} + \frac{12x}{L_i^3} \right] u_{2z} + \left[-\frac{2}{L_i} + \frac{6x}{L_i^2} \right] \varphi_{2y} \right) \end{aligned}, \tag{3.854}$$

$$Q_z^{\mathrm{e}}(x) = EI_y \left(\left[-\frac{12}{L_i^3} \right] u_{1z} + \left[+\frac{6}{L_i^2} \right] \varphi_{1y} + \left[+\frac{12}{L_i^3} \right] u_{2z} + \left[+\frac{6}{L_i^2} \right] \varphi_{2y} \right). \tag{3.855}$$

Thus, the normal and shear stress distributions can generally be calculated as follows:

$$\sigma_x^{\mathrm{e}}(x, z) = \frac{M_y^{\mathrm{e}}(x)}{I_y} \times z, \tag{3.856}$$

$$\tau_{xz}^{\mathrm{e}}(x, z) = \frac{Q_z^{\mathrm{e}}(x)}{2I_y} \left[\left(\frac{b}{2} \right)^2 - z^2 \right]. \tag{3.857}$$

The nodal values of the internal reactions for calculating the normal and shear stresses are summarized in Table 3.5. It is noted here that in this case the nodal values based on the finite element calculation are identical to the analytical solution.

From Table 3.5 it can be concluded that the maximum normal stress in each element is reached at the right node and that the shear stress is constant in each element. Thus, the critical stresses in each element can be calculated as follows:

$$\sigma_{x,\mathrm{I}} = -\frac{F_0 L}{a\, b_{\mathrm{I}}{}^2}, \tag{3.858}$$

$$\sigma_{x,\mathrm{II}} = -\frac{3F_0 L}{2a\, b_{\mathrm{II}}{}^2}, \tag{3.859}$$

Table 3.5 Nodal values of the internal reactions bending moment (M_y) and shear force (Q_z) at each node

	Element I		Element II	
	Left node	Right node	Left node	Right node
M_y	0	$-\frac{F_0 L}{6}$	$-\frac{F_0 L}{6}$	$-\frac{F_0 L}{4}$
Q_z	$-\frac{F_0}{2}$	$-\frac{F_0}{2}$	$-\frac{F_0}{2}$	$-\frac{F_0}{2}$

or for the shear stresses:

$$\tau_{xz,\mathrm{I}} = -\frac{3F_0}{4a\,b_\mathrm{I}}, \tag{3.860}$$

$$\tau_{xz,\mathrm{II}} = -\frac{3F_0}{4a\,b_\mathrm{II}}. \tag{3.861}$$

The objective function of the optimization problem, i.e. the mass of the stepped beam, can be represented as a function of the two design variables $X_1 = b_\mathrm{I}$ and $X_2 = b_\mathrm{II}$ as follows:

$$F(X_1, X_2) = \varrho L_\mathrm{I} a X_1 + \varrho L_\mathrm{II} a X_2 = \varrho L a \left(\frac{X_1}{3} + \frac{X_2}{6} \right), \tag{3.862}$$

where the minimization is to be carried out under consideration of the following seven inequality constraints:

$$g_1(X_1, X_2) = +\frac{F_0 L^3 (19 X_1^3 + 8 X_2^3)}{108 E a X_1^3 X_2^3 r_1 L} - 1 \le 0 \quad \text{(max. deflection)}, \tag{3.863}$$

$$g_2(X_1, X_2) = \frac{F_0 L}{a X_1^2} - R_{\mathrm{p}0.2} \le 0 \quad \text{(normal stress in I)}, \tag{3.864}$$

$$g_3(X_1, X_2) = \frac{3 F_0 L}{2a X_2^2} - R_{\mathrm{p}0.2} \le 0 \quad \text{(normal stress in II)}, \tag{3.865}$$

$$g_4(X_1, X_2) = \frac{3F_0}{4a X_1} - \frac{R_{\mathrm{p}0.2}}{2} \le 0 \quad \text{(shear stress in I)}, \tag{3.866}$$

$$g_5(X_1, X_2) = \frac{3F_0}{4a X_2} - \frac{R_{\mathrm{p}0.2}}{2} \le 0 \quad \text{(shear stress in II)}, \tag{3.867}$$

$$g_6(X_1, X_2) = X_1 - 20a \le 0 \quad \text{(height-to-width ratio in I)}, \tag{3.868}$$

$$g_7(X_1, X_2) = X_2 - 20a \le 0 \quad \text{(height-to-width ratio in II)}. \tag{3.869}$$

The graphical representation of the objective function and the inequality constraints (g_1, g_2, g_3) in the X_1-X_2 design space is shown in Fig. 3.45.

From Fig. 3.45 it can be concluded that considering g_1, g_2, and g_3 can be sufficient if a reasonable range of design variables is chosen. Furthermore, from Fig. 3.45 it can be concluded that the minimum is achieved for the case that g_1 is tangent to the objective function F. In order to derive an analytical solution for the minimum, one must take into account that the objective function F, in the representation $X_2(X_1) = \frac{6c}{\varrho L a} - 2X_1$, represents straight lines with a slope of –2. Thus, the condition for the minimum is that the curve g_1 assumes a slope of –2. Based on Eq. (3.863), the expression for g_1 can be rewritten:

$$X_2 = \left(-\frac{19 F_0 L^3 X_1^3}{8 F_0 L^3 - 108 E a r_1 L X_1^3} \right)^{\frac{1}{3}}, \tag{3.870}$$

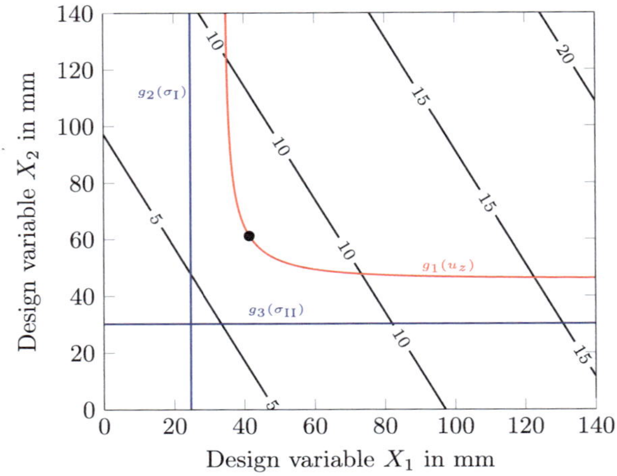

Fig. 3.45 Optimal design of a simply supported beam with three sections (consideration of symmetry): $L = 2540$ mm, $a = 45.16$ mm, $r_1 = 0.01$, $F_0 = 2667$ N, $E = 68948$ MPa, $R_{p0.2} = 247$ MPa, $\varrho = 2.691 \times 10^{-6}$ kg/mm^3, (original set of Eqs. (3.862)–(3.869)). Exact solution for the minimum: $X_{1,\mathrm{extr}} = 41.437$ mm, $X_{2,\mathrm{extr}} = 61.173$ mm (indicated by the • marker)

which, after short calculation, yields the expression of the first-order derivative as follows:

$$\left.\frac{\mathrm{d}X_2}{\mathrm{d}X_1}\right|_{g_1} = \frac{219^{\frac{1}{3}}\, F^{\frac{4}{3}}\, L^{\frac{8}{3}}}{\left(8\,F\,L^2 - 108Ea\,r_1X_1{}^3\right)^{\frac{1}{3}}\left(27Ea\,r_1X_1{}^3 - 2\,F\,L^2\right)} \stackrel{!}{=} -2. \tag{3.871}$$

The last equation can be rearranged so that one side has a root, and Newton's method (see Sect. A.1.3) can be used to determine the root. It is important that the root is sought in a reasonable interval, e.g. $40 \leq X_{1,\,\mathrm{extr}} \leq 60$.

3.4 Sandwich Structures: Basic Mechanical Load Cases

2.4.44 **Average bending stiffness of a sandwich**

The generalization of Eq. $(5.1)_\mathrm{TB}$ for different layer widths b^k results in:

$$\overline{EI}_y = \sum_{k=1}^{3} E^k \left[\tfrac{1}{12} b^k \left(\Delta h^k\right)^3 + b^k\,\Delta h^k \left(z_\mathrm{c}^k\right)^2\right]. \tag{3.872}$$

Application to our three-layer problem gives:

$$\overline{EI}_y = \frac{E^\mathrm{F} b^\mathrm{F} (\Delta h^\mathrm{F})^3}{6} + \frac{E^\mathrm{F} b^\mathrm{F} \Delta h^\mathrm{F} (\Delta h^\mathrm{C} + \Delta h^\mathrm{F})^2}{2} + \frac{E^\mathrm{C} b^\mathrm{C} (\Delta h^\mathrm{C})^3}{12}. \tag{3.873}$$

For the special case $b^\mathrm{C} = \frac{b^\mathrm{F}}{3}$ and $\Delta h^\mathrm{F} = \frac{\Delta h^\mathrm{C}}{4}$ the simplified formula results in:

$$\overline{EI}_y = \frac{b^\mathrm{F} (\Delta h^\mathrm{F})^3}{9} \times \left[114 E^\mathrm{F} + 16 E^\mathrm{C}\right]. \tag{3.874}$$

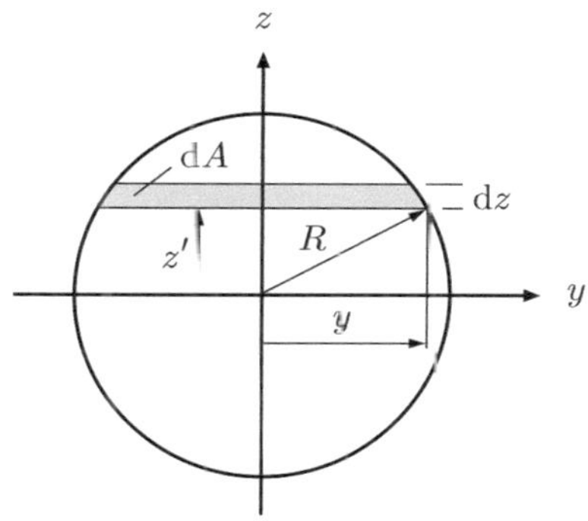

Fig. 3.46 Circular cross section for deriving the shear stress distribution τ_{xz}

2.4.45 **Calculation of the shear stress distribution for a circular cross section**

The starting point is again an infinitesimal beam element as in Fig. A.2. However, now the cross section is as in Fig. 3.46.

The forces equilibrium in the x direction results in, see Fig. A.3:

$$-\sigma_x(x)\underbrace{b(z)\mathrm{d}z}_{\mathrm{d}A}+\sigma_x(x+\mathrm{d}x)\underbrace{b(z)\mathrm{d}z}_{\mathrm{d}A}-\tau_{zx}(z)b(z)\mathrm{d}x=0, \tag{3.875}$$

or simplified after a Taylor series expansion of the stress at $(x+\mathrm{d}x)$:

$$\frac{\mathrm{d}\sigma_x(x)}{\mathrm{d}x}\mathrm{d}x\mathrm{d}A-\tau_{zx}(z)b(z)\mathrm{d}x=0, \tag{3.876}$$

or

$$\frac{\mathrm{d}\sigma_x(x)}{\mathrm{d}x}\mathrm{d}A-\tau_{zx}(z)b(z)=0. \tag{3.877}$$

Converted to

$$\tau_{zx}(z)=\frac{\mathrm{d}\sigma_x(x)}{b(z)\mathrm{d}x}\mathrm{d}A, \tag{3.878}$$

one obtains with

$$\frac{\mathrm{d}\sigma_x(x)}{\mathrm{d}x}=\frac{\mathrm{d}}{\mathrm{d}x}\left(\frac{M_y(x)}{I_y}\times z\right)=\frac{z}{I_y}\times\frac{\mathrm{d}M_y(x)}{\mathrm{d}x}=\frac{Q_z(x)}{I_y}\times z \tag{3.879}$$

the shear stress distribution:

$$\tau_{zx}(z)=\frac{1}{b(z)}\int\frac{Q_z(x)}{I_y}\times z'\mathrm{d}A=\frac{Q_z(x)}{I_yb(z)}\int z'\mathrm{d}A=\frac{Q_z(x)\mathcal{H}_y(z)}{I_yb(z)}, \tag{3.880}$$

where $\mathcal{H}_y(z)$ represents the axial first moment of area for the part of the cross section shown in Fig. A.3. Assuming a circular cross section with radius R, this moment, taking into account $\mathrm{d}A = 2y\mathrm{d}z$ and $y = \sqrt{R^2 - z^2}$, is:

$$\mathcal{H}_y(z) = \int z'\mathrm{d}A = \int_z^R z' 2y\mathrm{d}z' = 2\int_z^R z'\sqrt{R^2 - z'^2}\,\mathrm{d}z' \tag{3.881}$$

$$= 2\left[-\frac{1}{3}\left(R^2 - z'^2\right)^{\frac{3}{2}}\right]_z^R = 2\left(+\frac{1}{3}\left(R^2 - z^2\right)^{\frac{3}{2}}\right) \tag{3.882}$$

$$= \frac{2}{3}\left(R^2 - z^2\right)^{\frac{3}{2}}. \tag{3.883}$$

If one further considers that $b(z) = 2y = 2(R^2 - z^2)^{\frac{1}{2}}$, the shear stress distribution using $\tau_{zx}(z) = \tau_{xz}(z)$ is:

$$\tau_{xz}(z) = \frac{Q_z(x)}{3I_y}\left(R^2 - z^2\right). \tag{3.884}$$

The maximum shear stress is given for $z = 0$:

$$\tau_{xz,\max} = \frac{Q_z(x)R^2}{3I_y} = \frac{4Q_z(x)}{3\pi R^2} = \frac{4Q_z(x)}{3A}. \tag{3.885}$$

2.4.46 **Limit case of the normalized stiffness of a sandwich**

From Eq. (2.5) it follows using $E^{\mathrm{F}} = E^{\mathrm{C}} = E$ and $\Delta h^{\mathrm{F}} = \Delta h^{\mathrm{C}} = \frac{1}{3}$:

$$\frac{\overline{EI}_y}{\overline{EI}_{y,\mathrm{ref}}} = \left[\frac{1}{4} + \frac{3}{4}4 + \frac{1}{8}\right] = \frac{27}{8}. \tag{3.886}$$

Sandwich of height '1': $\overline{EI}_y = \frac{1}{12}b1^3 = \frac{b}{12}$; reference configuration of height '$\frac{2}{3}$': $\overline{EI}_{y,\mathrm{ref}} = \frac{1}{12}b\left(\frac{2}{3}\right)^3 = \frac{2b}{81}$. Ratio: $\frac{\overline{EI}_y}{\overline{EI}_{y,\mathrm{ref}}} = \frac{81}{2\times 12} = \frac{27}{8}$.

2.4.47 **Deflection of a sandwich beam with distributed load using the partial deflection method**

- Internal reactions (Q_z, M_y):

The functions of the internal reactions can be found in the solution of Exercise 2.2.12:

$$Q_z(x) = q_0\left(x - \frac{L}{2}\right), \tag{3.887}$$

$$M_y(x) = \frac{q_0}{2}\left(x - \frac{L}{2}\right). \tag{3.888}$$

• General solution:

The bending and shear differential equations can be taken from Eqs. (5.93)$_{\text{TB}}$ and (5.104)$_{\text{TB}}$:

$$\frac{\mathrm{d}^2}{\mathrm{d}x^2}\left(\overline{EI_y}\frac{\mathrm{d}^2 u_{z,\mathrm{b}}(x)}{\mathrm{d}x^2}\right) = q_z(x), \tag{3.889}$$

$$\frac{\mathrm{d}u_{z,\mathrm{s}}}{\mathrm{d}x} = \frac{Q_z(x)}{AG^{\mathrm{C}}}. \tag{3.890}$$

With $q_z(x) = -q_0$ and $\overline{EI_y} = \text{const.}$ respectively $AG^{\mathrm{C}} = \text{const.}$, integration gives:

$$\overline{EI_y}\frac{\mathrm{d}^4 u_{z,\mathrm{b}}(x)}{\mathrm{d}x^4} = -q_0, \tag{3.891}$$

$$\overline{EI_y}\frac{\mathrm{d}^3 u_{z,\mathrm{b}}(x)}{\mathrm{d}x^3} = -q_0 x + c_1 = -Q_z(x), \tag{3.892}$$

$$\overline{EI_y}\frac{\mathrm{d}^2 u_{z,\mathrm{b}}(x)}{\mathrm{d}x^2} = -\frac{1}{2}q_0 x^2 + c_1 x + c_2 = -M_y(x), \tag{3.893}$$

$$\overline{EI_y}\frac{\mathrm{d}^1 u_{z,\mathrm{b}}(x)}{\mathrm{d}x^1} = -\frac{1}{6}q_0 x^3 + \frac{1}{2}c_1 x^2 + c_2 x + c_3 = -\overline{EI_y}\varphi_y(x), \tag{3.894}$$

$$\overline{EI_y}u_{z,\mathrm{b}}(x) = -\frac{1}{24}q_0 x^4 + \frac{1}{6}c_1 x^3 + \frac{1}{2}c_2 x^2 + c_3 x + c_4, \tag{3.895}$$

respectively

$$\frac{\mathrm{d}u_{z,\mathrm{s}}}{\mathrm{d}x} = \frac{q_0\left(x - \frac{L}{2}\right)}{AG^{\mathrm{C}}}, \tag{3.896}$$

$$u_{z,\mathrm{s}} = \frac{q_0}{AG^{\mathrm{C}}}\left(\frac{x^2}{2} - \frac{L}{2}x\right) + c_5. \tag{3.897}$$

• Particular solution:

Using the boundary conditions $u_{z,\mathrm{b}}(0) = u_{z,\mathrm{s}}(0) = u_{z,\mathrm{b}}(L) = 0$ and $M_y(0) = M_y(L) = 0$ the constants of integration result in $c_2 = c_4 = c_5 = 0$, $c_1 = \frac{q_0 L}{2}$ and $c_3 = -\frac{q_0 L^3}{24}$. Thus, the particular solution results from superposition of the bending and shear components to:

$$u_z(x) = -\frac{q_0 L^4}{24\overline{EI_y}}\left(\left(\frac{x}{L}\right)^4 - 2\left(\frac{x}{L}\right)^3 + \left(\frac{x}{L}\right)\right) - \frac{q_0 L^2}{2AG^{\mathrm{C}}}\left(-\left(\frac{x}{L}\right)^2 + \left(\frac{x}{L}\right)\right). \tag{3.898}$$

• Maximum value of deflection:

The maximum deflection is reached in the middle of the beam:

$$u_z(\tfrac{L}{2}) = -\frac{5q_0L^4}{384\overline{EI}_y} - \frac{q_0L^2}{8AG^{\mathrm{C}}}. \tag{3.899}$$

• Ratio of the partial deflections:

$$\frac{u_{z,\mathrm{b}}(\frac{L}{2})}{u_{z,\mathrm{s}}(\frac{L}{2})} = \frac{5}{24} \times \frac{L^2G^{\mathrm{C}}}{E^{\mathrm{F}}\Delta h^{\mathrm{C}}\Delta h^{\mathrm{F}}} = 0.165 = 16.5\%. \tag{3.900}$$

2.4.48 **Deflection of a sandwich beam (cantilever) with distributed load using the partial deflection method**

• Internal reactions (Q_z, M_y):

The functions of the internal reactions can be taken from the solution of Exercise 2.2.11:

$$Q_z(x) = -q_0\,(L-x)\,, \tag{3.901}$$

$$M_y(x) = \frac{q_0}{2}\,(L-x)^2\,. \tag{3.902}$$

• General solution:

The bending and shear differential equations can be taken from Eqs. $(5.93)_{\mathrm{TB}}$ and $(5.104)_{\mathrm{TB}}$:

$$\frac{\mathrm{d}^2}{\mathrm{d}x^2}\left(\overline{EI}_y\frac{\mathrm{d}^2u_{z,\mathrm{b}}(x)}{\mathrm{d}x^2}\right) = q_z(x), \tag{3.903}$$

$$\frac{\mathrm{d}u_{z,\mathrm{s}}}{\mathrm{d}x} = \frac{Q_z(x)}{AG^{\mathrm{C}}}. \tag{3.904}$$

With $q_z(x) = -q_0$ and $\overline{EI}_y = \text{const.}$ respectively $AG^{\mathrm{C}} = \text{const.}$, integration gives:

$$\overline{EI}_y\frac{\mathrm{d}^4u_{z,\mathrm{b}}(x)}{\mathrm{d}x^4} = -q_0, \tag{3.905}$$

$$\overline{EI}_y\frac{\mathrm{d}^3u_{z,\mathrm{b}}(x)}{\mathrm{d}x^3} = -q_0x + c_1 = -Q_z(x), \tag{3.906}$$

$$\overline{EI}_y\frac{\mathrm{d}^2u_{z,\mathrm{b}}(x)}{\mathrm{d}x^2} = -\frac{1}{2}q_0x^2 + c_1x + c_2 = -M_y(x), \tag{3.907}$$

$$\overline{EI}_y\frac{\mathrm{d}^1u_{z,\mathrm{b}}(x)}{\mathrm{d}x^1} = -\frac{1}{6}q_0x^3 + \frac{1}{2}c_1x^2 + c_2x + c_3 = -\overline{EI}_y\varphi_y(x), \tag{3.908}$$

$$\overline{EI}_y u_{z,\mathrm{b}}(x) = -\frac{1}{24} q_0 x^4 + \frac{1}{6} c_1 x^3 + \frac{1}{2} c_2 x^2 + c_3 x + c_4, \tag{3.909}$$

respectively

$$\frac{\mathrm{d}u_{z,\mathrm{s}}}{\mathrm{d}x} = \frac{-q_0\,(L-x)}{AG^{\mathrm{C}}}, \tag{3.910}$$

$$u_{z,\mathrm{s}} = \frac{q_0}{AG^{\mathrm{C}}}\left(\frac{x^2}{2} - Lx\right) + c_5. \tag{3.911}$$

- Particular solution:

Using the boundary conditions $u_{z,\mathrm{b}}(0) = u_{z,\mathrm{s}}(0) = 0$, $\frac{\mathrm{d}u_{z,\mathrm{b}}(0)}{\mathrm{d}x} = 0$, $M_y(L) = 0$ and $Q_z(L) = 0$ the constants of integration result in $c_3 = c_4 = c_5 = 0$, $c_1 = q_0 L$ and $c_2 = -\frac{1}{2} q_0 L^2$. This gives the particular solution:

$$u_z(x) = -\frac{q_0 L^4}{2\overline{EI}_y}\left(\frac{1}{12}\left(\frac{x}{L}\right)^4 - \frac{1}{3}\left(\frac{x}{L}\right)^3 + \frac{1}{2}\left(\frac{x}{L}\right)^2\right) - \frac{q_0 L^2}{AG^{\mathrm{C}}}\left(-\frac{1}{2}\left(\frac{x}{L}\right)^2 + \left(\frac{x}{L}\right)\right). \tag{3.912}$$

- Maximum value of deflection:

The maximum deflection is reached at the right boundary of the beam:

$$u_z(L) = -\frac{q_0 L^4}{8\overline{EI}_y} - \frac{q_0 L^2}{2AG^{\mathrm{C}}}. \tag{3.913}$$

- Ratio of the partial deflections:

$$\frac{u_{z,\mathrm{b}}(L)}{u_{z,\mathrm{s}}(L)} = \frac{1}{4} \times \frac{L^2 G A^{\mathrm{C}}}{\overline{EI}_y} = \frac{1}{2} \times \frac{L^2 G^{\mathrm{C}}}{E^{\mathrm{F}} \Delta h^{\mathrm{C}} \Delta h^{\mathrm{F}}} = 0.397 = 39.7\%. \tag{3.914}$$

2.4.49 **Deflection of a sandwich beam (cantilever) with shear force using the partial deflection method**

- Internal reactions (Q_z, M_y):

The functions of the internal reactions can be found in the solution of Exercise 2.1.2:

$$Q_z(x) = -F_0, \tag{3.915}$$

$$M_y(x) = F_0\,(L-x). \tag{3.916}$$

The internal reactions for the problem according to Fig. 2.39 are shown in Fig. 3.47.

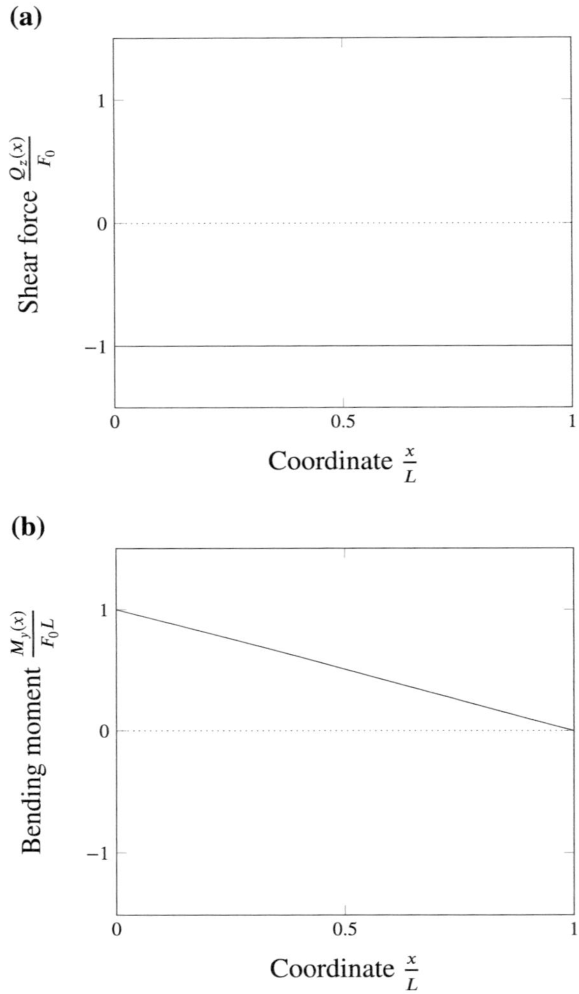

Fig. 3.47 Internal reactions for a sandwich beam according to Fig. 2.39, i.e. cantilever with end shear force F_0 at $x = L$ in negative z-direction: **a** shear force and **b** bending moment diagram

- General solution:

The bending and shear differential equations can be taken from Eqs. $(5.95)_{\text{TB}}$ and $(5.104)_{\text{TB}}$:

$$\overline{EI_y}\frac{\mathrm{d}^2 u_{z,\,\mathrm{b}}(x)}{\mathrm{d}x^2} = -M_y(x), \tag{3.917}$$

$$\frac{\mathrm{d}u_{z,\,\mathrm{s}}}{\mathrm{d}x} = \frac{Q_z(x)}{AG^{\mathrm{C}}}. \tag{3.918}$$

With $q_z(x) = 0$ and $\overline{EI_y} = \text{const.}$ respectively $AG^{\text{C}} = \text{const.}$, integration gives:

$$\overline{EI_y}\frac{\mathrm{d}^2 u_{z,\text{b}}(x)}{\mathrm{d}x^2} = -M_y(x) = -F_0(L-x), \tag{3.919}$$

$$\overline{EI_y}\frac{\mathrm{d}^1 u_{z,\text{b}}(x)}{\mathrm{d}x^1} = -F_0\left(Lx - \frac{x^2}{2}\right) + c_1 = -\overline{EI_y}\varphi_y(x), \tag{3.920}$$

$$\overline{EI_y}u_{z,\text{b}}(x) = -F_0\left(L\frac{x^2}{2} - \frac{x^3}{6}\right) + c_1 x + c_2, \tag{3.921}$$

respectively

$$\frac{\mathrm{d}u_{z,\text{s}}}{\mathrm{d}x} = \frac{-F_0}{AG^{\text{C}}}, \tag{3.922}$$

$$u_{z,\text{s}} = \frac{-F_0}{AG^{\text{C}}}x + c_3. \tag{3.923}$$

- Particular solution:

Using the boundary conditions $u_{z,\text{b}}(0) = u_{z,\text{s}}(0) = 0$ and $\frac{\mathrm{d}u_{z,\text{b}}(0)}{\mathrm{d}x} = 0$ the constants of integration result in $c_1 = c_2 = c_3 = 0$. This gives the particular solution:

$$u_z(x) = -\frac{F_0 L^3}{6\overline{EI_y}}\left(3\left(\frac{x}{L}\right)^2 - \left(\frac{x}{L}\right)^3\right) - \frac{F_0 L}{AG^{\text{C}}}\left(\frac{x}{L}\right). \tag{3.924}$$

- Maximum value of deflection:

The maximum deflection is reached at the right boundary of the beam:

$$u_z(L) = -\frac{F_0 L^3}{3\overline{EI_y}} - \frac{F_0 L}{AG^{\text{C}}}. \tag{3.925}$$

- Ratio of the partial deflections:

$$\frac{u_{z,\text{b}}(L)}{u_{z,\text{s}}(L)} = \frac{2}{3} \times \frac{L^2 G^{\text{C}}}{E^{\text{F}}\Delta h^{\text{C}}\Delta h^{\text{F}}} = 0.529 = 52.9\%. \tag{3.926}$$

2.4.50 **Deflection of a sandwich beam (cantilever) with shear force and bending moment using the partial deflection method**

• Internal reactions (Q_z, M_y):

The functions of the internal reactions can be adapted from the solution of the Exercise 2.3.31:

$$Q_z(x) = -F_0, \tag{3.927}$$

$$M_y(x) = F_0 L \left(\frac{3}{2} - \frac{x}{L} \right). \tag{3.928}$$

• General solution:

The bending and shear differential equations can be taken from Eqs. $(5.94)_{\text{TB}}$ and $(5.104)_{\text{TB}}$:

$$\frac{\mathrm{d}}{\mathrm{d}x}\left(\overline{EI_y} \frac{\mathrm{d}^2 u_{z,\mathrm{b}}(x)}{\mathrm{d}x^2} \right) = -Q_z(x), \tag{3.929}$$

$$\frac{\mathrm{d}u_{z,\mathrm{s}}}{\mathrm{d}x} = \frac{Q_z(x)}{AG^{\mathrm{C}}}. \tag{3.930}$$

With $q_z(x) = 0$ and $\overline{EI_y} = \text{const.}$ or $AG^{\mathrm{C}} = \text{const.}$, integration gives:

$$\overline{EI_y} \frac{\mathrm{d}^3 u_{z,\mathrm{b}}(x)}{\mathrm{d}x^3} = -Q_z(x) = F_0, \tag{3.931}$$

$$\overline{EI_y} \frac{\mathrm{d}^2 u_{z,\mathrm{b}}(x)}{\mathrm{d}x^2} = F_0 x + c_1 = -M_y(x), \tag{3.932}$$

$$\overline{EI_y} \frac{\mathrm{d}^1 u_{z,\mathrm{b}}(x)}{\mathrm{d}x^1} = \frac{F_0}{2} x^2 + c_1 x + c_2 = -\overline{EI_y} \varphi_y(x), \tag{3.933}$$

$$\overline{EI_y} u_{z,\mathrm{b}} = \frac{F_0}{6} x^3 + \frac{c_1}{2} x^2 + c_2 x + c_3, \tag{3.934}$$

respectively

$$\frac{\mathrm{d}u_{z,\mathrm{s}}}{\mathrm{d}x} = \frac{-F_0}{AG^{\mathrm{C}}}, \tag{3.935}$$

$$u_{z,\mathrm{s}} = -\frac{F_0}{AG^{\mathrm{C}}} x + c_4. \tag{3.936}$$

Using the boundary conditions $u_{z,\mathrm{b}}(0) = u_{z,\mathrm{s}}(0) = 0$, $\frac{\mathrm{d}u_{z,\mathrm{b}}(0)}{\mathrm{d}x} = 0$ and $M_y(L) = \frac{F_0 L}{2}$ the constants of integration result in $c_1 = -\frac{3F_0L}{2}$ and $c_2 = c_3 = c_4 = 0$. This gives the particular solution:

$$u_z(x) = \frac{1}{\overline{EI}_y}\left(\frac{F_0x^2}{6} - \frac{3F_0Lx^2}{4}\right) - \frac{F_0x}{AG^{\mathrm{C}}}. \tag{3.937}$$

• Maximum value of deflection:

The maximum deflection is reached at the right boundary of the beam:

$$u_z(L) = -\frac{7F_0L^3}{12\overline{EI}_y} - \frac{F_0L}{AG^{\mathrm{C}}}. \tag{3.938}$$

• Ratio of the partial deflections:

$$\frac{u_{z,\mathrm{s}}(L)}{u_{z,\mathrm{b}}(L)} = \frac{6}{7} \times \frac{E^{\mathrm{F}}\Delta h^{\mathrm{F}}\Delta h_{\mathrm{c}}}{L^2G^{\mathrm{C}}} = 1.12 = 112\%. \tag{3.939}$$

3.5 Sandwich Structures: Limit Load

2.5.51 **Failure analysis of a sandwich beam with distributed load**

• Internal reactions (Q_z, M_y):

The functions of the internal reactions can be taken from the solution of Exercise 2.2.12:

$$Q_z(x) = q_0L\left(\left[\frac{x}{L}\right] - \frac{1}{2}\right), \tag{3.940}$$

$$M_y(x) = \frac{q_0L^2}{2}\left(\left[\frac{x}{L}\right]^2 - \left[\frac{x}{L}\right]^1\right) < 0, \tag{3.941}$$

where the maximum absolute values are as follows:

$$|Q_{z,\max}| = \frac{q_0L}{2}, \tag{3.942}$$

$$|M_{y,\max}| = \frac{q_0L^2}{8}. \tag{3.943}$$

The courses of the internal reactions and the corresponding maximum values are shown in Fig. 3.48.

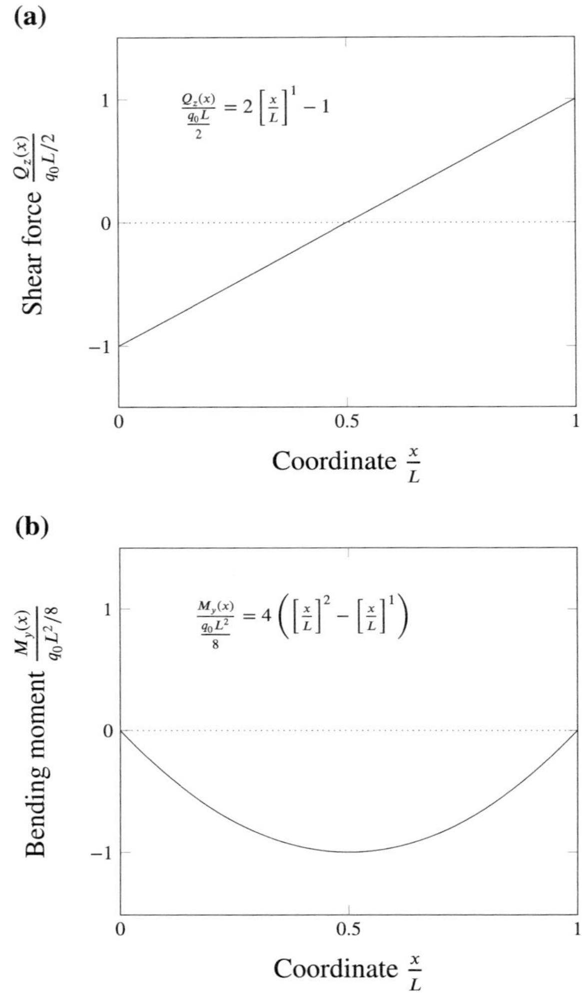

Fig. 3.48 Internal reactions for a sandwich beam with distributed load (configuration: hinged on both sides with constant distributed load q_0 in negative z-direction): **a** shear force and **b** bending moment diagram

- Theory

Checking whether one of the simplifying theories is applicable:

$$\frac{6E^{\mathrm{F}}\Delta h^{\mathrm{F}}(h_{\mathrm{c}})^2}{E^{\mathrm{C}}(\Delta h^{\mathrm{C}})^3} = 526.77 > 100, \tag{3.944}$$

$$\frac{\Delta h^{\mathrm{C}}}{\Delta h^{\mathrm{F}}} = 30.0 > 4.77. \tag{3.945}$$

Thus, the theory for soft cores and thin face sheets can be applied.

• Maximum normal stress in the face sheets:

Using Eq. (5.49)$_{\text{TB}}$, i.e.

$$\sigma_{x,\text{F}} = \frac{|M_{y,\max}|}{b\Delta h^{\text{F}}\Delta h^{\text{C}}}, \tag{3.946}$$

the maximum normal stress in the face sheets with the given numerical values is:

$$\sigma_{x,\text{F}} = 8.33\,\text{MPa} < R^{\text{F}}_{\text{p0.2}}. \tag{3.947}$$

Thus, there is no failure in the face sheets.

• Maximum shear stress in the core:

Using Eq. (5.91)$_{\text{TB}}$, i.e.

$$\tau_{zx,\text{C}} = \frac{|Q_{z,\max}|}{bh_{\text{c}}}, \tag{3.948}$$

the maximum shear stress in the core with the given numerical values is:

$$\tau_{zx,\text{C}} = 0.081\,\text{MPa} < \tau^{\text{C}}_{\text{aB}}. \tag{3.949}$$

Thus, there is no failure in the core.

• Shear stress in the adhesive layer:

The shear stress in the adhesive layer is in this case equal to the constant shear stress in the core:

$$\tau_{zx} = 0.081\,\text{MPa} < \tau_{\text{aB}}. \tag{3.950}$$

Thus, there is no failure in the adhesive layer.

• Local wrinkling of the compression face sheet:

The critical stress is generally calculated according to Eq. (6.28)$_{\text{TB}}$ as:

$$\sigma_{\text{cr}} = B_1 \left(E^{\text{F}}\right)^{\frac{1}{3}} \left(E^{\text{C}}\right)^{\frac{2}{3}}. \tag{3.951}$$

The factor B_1 can be numerically approximated using the Python program B1.py (see Appendix A.2). With $B_1 = 0.567$, the critical stress is:

$$\sigma_{\text{cr}} = 229.904\,\text{MPa} > \sigma_{x,\text{F}}. \tag{3.952}$$

Thus, there is no failure due to local wrinkling of the compression face sheet.

2.5.52 **Failure analysis of a sandwich beam (cantilever) with distributed load**

• Internal reactions (Q_z, M_y):

The functions of the internal reactions can be taken from the solution of Exercise 2.1.2b:

$$Q_z(x) = q_0 L \left(\left[\frac{x}{L} \right]^1 - 1 \right), \tag{3.953}$$

$$M_y(x) = \frac{q_0 L^2}{2} \left(\left[\frac{x}{L} \right]^2 - 2 \left[\frac{x}{L} \right]^1 + 1 \right) > 0, \tag{3.954}$$

where the maximum values are as follows:

$$Q_{z,\text{max}} = q_0 L, \tag{3.955}$$

$$M_{y,\text{max}} = \frac{q_0 L^2}{2}. \tag{3.956}$$

The courses of the internal reactions and the corresponding maximum values are shown in Fig. 3.49.

• Theory

Checking whether one of the simplifying theories is applicable:

$$\frac{6 E^{\text{F}} \Delta h^{\text{F}} (h_{\text{c}})^2}{E^{\text{C}} (\Delta h^{\text{C}})^3} = 526.77 > 100, \tag{3.957}$$

$$\frac{\Delta h^{\text{C}}}{\Delta h^{\text{F}}} = 30.0 > 4.77. \tag{3.958}$$

Thus, the theory for soft cores and thin face sheets can be applied.

• Maximum normal stress in the face sheets:

Using Eq. $(5.49)_{\text{TB}}$, i.e.

$$\sigma_{x,\text{F}} = \frac{|M_{y,\text{max}}|}{b \Delta h^{\text{F}} \Delta h^{\text{C}}}, \tag{3.959}$$

the maximum normal stress in the face sheets with the given numerical values is:

$$\sigma_{x,\text{F}} = 33.22\,\text{MPa} < R^{\text{F}}_{\text{p}0.2}. \tag{3.960}$$

Thus, there is no failure in the face sheets.

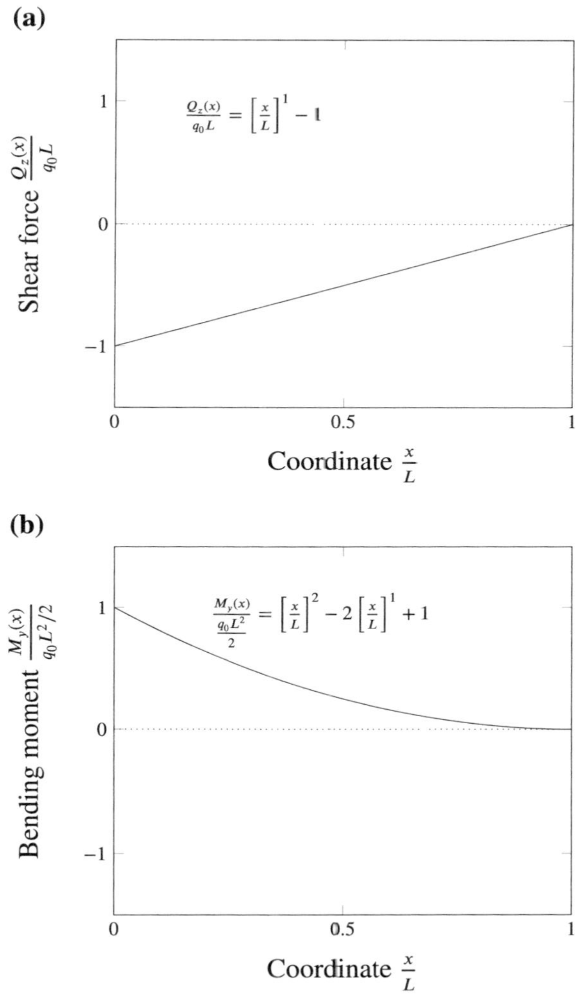

Fig. 3.49 Internal reactions for a sandwich beam with distributed load (configuration: cantilever beam with constant distributed load q_0 in negative z-direction): **a** shear force and **b** bending moment diagram

• Maximum shear stress in the core:

Using Eq. $(5.91)_{\text{TB}}$, i.e.

$$\tau_{zx,\text{C}} = \frac{|Q_{z,\max}|}{bh_\text{c}}, \tag{3.961}$$

the maximum shear stress in the core with the given numerical values is:

$$\tau_{zx,\text{C}} = 0.163\,\text{MPa} < \tau_{\text{aB}}^{\text{C}}. \tag{3.962}$$

Thus, there is no failure in the core.

• Shear stress in the adhesive layer:

The shear stress in the adhesive layer is in this case equal to the constant shear stress in the core:

$$\tau_{zx} = 0.163\,\text{MPa} < \tau_{\text{aB}}. \tag{3.963}$$

Thus, there is no failure in the adhesive layer.

• Local wrinkling of the compression face sheet:

The critical stress is generally calculated according to Eq. $(6.28)_{\text{TB}}$ as:

$$\sigma_{\text{cr}} = B_1 \left(E^{\text{F}}\right)^{\frac{1}{3}} \left(E^{\text{C}}\right)^{\frac{2}{3}}. \tag{3.964}$$

The factor B_1 can be numerically approximated using the Python program B1.py (see Appendix A.2). With $B_1 = 0.567$, the critical stress is:

$$\sigma_{\text{cr}} = 229.904\,\text{MPa} > \sigma_{x,\text{F}}. \tag{3.965}$$

Thus, there is no failure due to local wrinkling of the compression face sheet.

2.5.53 **Failure analysis of a sandwich beam under 4-point bending**

• Internal reactions (Q_z, M_y):

To determine the internal reactions, the inside must be cut free, see Fig. 3.50.

In the Section 1 with $0 \leq x \leq \frac{L}{4}$ we get:

$$\overset{\curvearrowright}{\sum} M_y = 0: \quad +F_0 x + M_y(x) = 0 \quad \rightarrow \quad M_y(x) = -F_0 x, \tag{3.966}$$

or for the vertical force equilibrium:

$$\overset{\uparrow}{\sum} F_z = 0: \quad +F_0 + Q_z(x) = 0 \quad \rightarrow \quad Q_z(x) = -F_0. \tag{3.967}$$

In the Section 2 with $\frac{L}{4} \leq x \leq \frac{3L}{4}$ we get:

$$\overset{\curvearrowright}{\sum} M_y = 0: \quad +F_0 x - F_0(x - \tfrac{L}{4}) + M_y(x) = 0 \quad \rightarrow \quad M_y(x) = -\frac{F_0 L}{4}, \tag{3.968}$$

or for the vertical force equilibrium:

$$\overset{\uparrow}{\sum} F_z = 0: \quad +F_0 - F_0 + Q_z(x) = 0 \quad \rightarrow \quad Q_z(x) = 0. \tag{3.969}$$

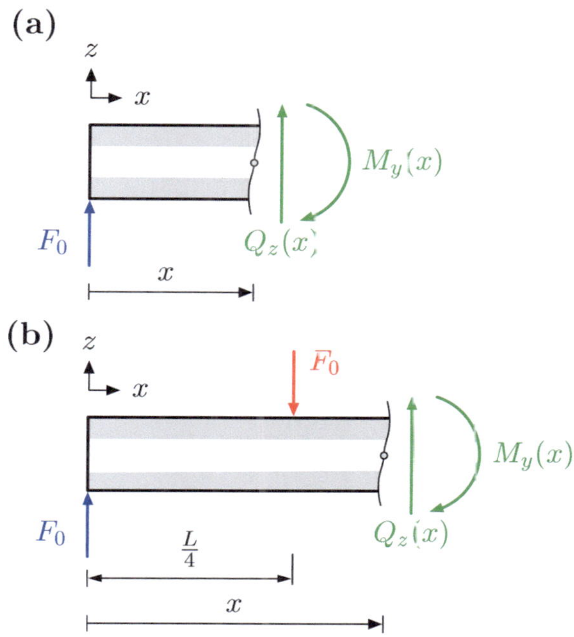

Fig. 3.50 Cutting free inside the beam under 4-point bending: **a** Section 1 with $0 \leq x \leq \frac{L}{4}$ and **b** Section 2 with $\frac{L}{4} \leq x \leq \frac{3L}{4}$

The internal reactions and the corresponding maximum values are shown in Fig. 3.51.
• Theory

Checking whether one of the simplifying theories is applicable:

$$\frac{6E^{\mathrm{F}}\Delta h^{\mathrm{F}}(h_{\mathrm{c}})^2}{E^{\mathrm{C}}(\Delta h^{\mathrm{C}})^3} = 71.04 < 100, \tag{3.970}$$

$$\frac{\Delta h^{\mathrm{C}}}{\Delta h^{\mathrm{F}}} = 1.0 < 4.77. \tag{3.971}$$

Thus, the exact theory must be applied.

• Maximum normal stress in the face sheets:

Using Eq. $(5.28)_{\mathrm{TB}}$, i.e.

$$\sigma_{x,\mathrm{F,max}} = \frac{\frac{F_0 L}{4} E^{\mathrm{F}}}{\frac{E^{\mathrm{F}} b(\Delta h^{\mathrm{F}})^3}{6} + \frac{E^{\mathrm{F}} b \Delta h^{\mathrm{F}}(h_{\mathrm{c}})^2}{2} + \frac{E^{\mathrm{C}} b(\Delta h^{\mathrm{C}})^3}{12}} \times \left(\Delta h^{\mathrm{F}} + \frac{\Delta h^{\mathrm{C}}}{2}\right), \tag{3.972}$$

the maximum normal stress in the face sheets with the given numerical values is:

$$\sigma_{x,\mathrm{F}} = 1.709\,\mathrm{MPa} < R^{\mathrm{F}}_{\mathrm{p0.2}}. \tag{3.973}$$

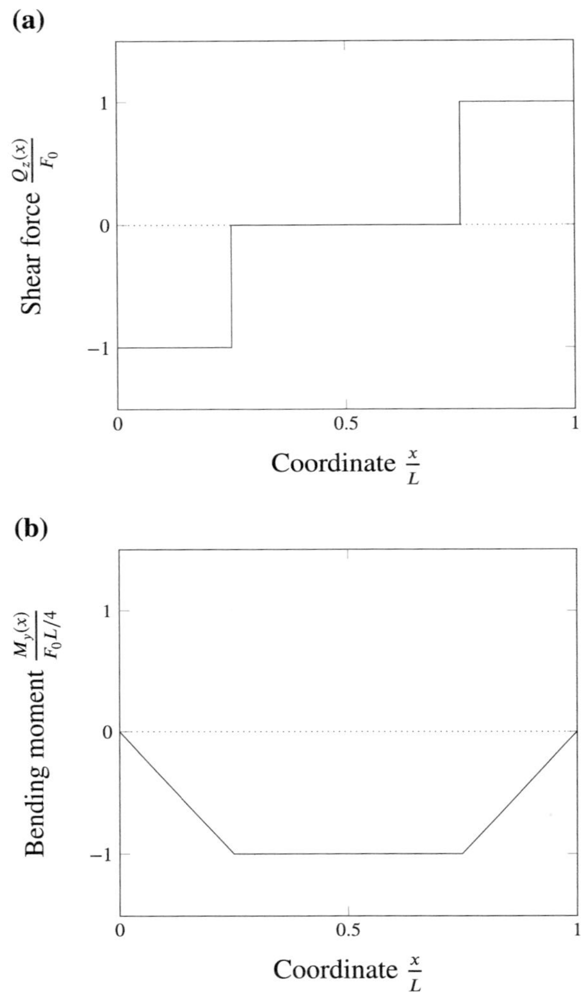

Fig. 3.51 Internal reactions for sandwich under 4-point bending: **a** shear force and **b** bending moment diagram

Thus, there is no failure in the face sheets.

- Maximum normal stress in the core:

Using Eq. (5.27)$_{\text{TB}}$, i.e.

$$\sigma_{x,\text{C,max}} = \frac{\frac{F_0 L}{4} E^{\text{C}}}{\dfrac{E^{\text{F}} b (\Delta h^{\text{F}})^3}{6} + \dfrac{E^{\text{F}} b \Delta h^{\text{F}} (h_{\text{c}})^2}{2} + \dfrac{E^{\text{C}} b (\Delta h^{\text{C}})^3}{12}} \times \left(\frac{\Delta h^{\text{C}}}{2} \right), \tag{3.974}$$

the maximum normal stress in the core with the given numerical values is:

$$\sigma_{x,\text{C}} = 0.192\,\text{MPa} < R_{\text{m}}^{\text{C}} \quad (< |\sigma_{\text{dB}}^{\text{C}}|). \tag{3.975}$$

Thus, there is no normal stress failure in the core.

- Maximum shear stress in the core:

Using Eq. (5.80)$_{\text{TB}}$, i.e.

$$\tau_{zx,\mathrm{C,max}} = \frac{|Q_{z,\mathrm{max}}|\left[E^{\mathrm{C}}\left(\left(\frac{\Delta h^{\mathrm{C}}}{2}\right)^2 - 0^2\right) + E^{\mathrm{F}}\left(\Delta h^{\mathrm{C}} + \Delta h^{\mathrm{F}}\right)\Delta h^{\mathrm{F}}\right]}{2\left(\frac{E^{\mathrm{F}}b(\Delta h^{\mathrm{F}})^3}{6} + \frac{E^{\mathrm{F}}b\Delta h^{\mathrm{F}}(h_{\mathrm{c}})^2}{2} + \frac{E^{\mathrm{C}}b(\Delta h^{\mathrm{C}})^3}{12}\right)}, \tag{3.976}$$

the maximum shear stress in the core with the given numerical values is:

$$\tau_{zx,\mathrm{C}} = 0.119\,\mathrm{MPa} < \tau_{\mathrm{aB}}^{\mathrm{C}}. \tag{3.977}$$

Thus, there is no shear failure in the core.

- Shear stress in the adhesive layer:

The shear stress in the adhesive layer is obtained in general, for example, by evaluating Eq. (5.80)$_{\text{TB}}$ at the location $z = \Delta h^{\mathrm{C}}/2$ as:

$$\tau_{zx} = \frac{|Q_{z,\mathrm{max}}|\left[E^{\mathrm{C}}\left(\left(\frac{\Delta h^{\mathrm{C}}}{2}\right)^2 - \left(\frac{\Delta h^{\mathrm{C}}}{2}\right)^2\right) + E^{\mathrm{F}}\left(\Delta h^{\mathrm{C}} + \Delta h^{\mathrm{F}}\right)\Delta h^{\mathrm{F}}\right]}{2\left(\frac{E^{\mathrm{F}}b(\Delta h^{\mathrm{F}})^3}{6} + \frac{E^{\mathrm{F}}b\Delta h^{\mathrm{F}}(h_{\mathrm{c}})^2}{2} + \frac{E^{\mathrm{C}}b(\Delta h^{\mathrm{C}})^3}{12}\right)}, \tag{3.978}$$

or with the given numerical values:

$$\tau_{zx} = 0.114\,\mathrm{MPa} < \tau_{\mathrm{aB}}. \tag{3.979}$$

Thus, there is no failure in the adhesive layer.

- Local wrinkling of the compression face sheet:

The critical stress is generally calculated according to Eq. (6.28)$_{\text{TB}}$ as:

$$\sigma_{\mathrm{cr}} = B_1\left(E^{\mathrm{F}}\right)^{\frac{1}{3}}\left(E^{\mathrm{C}}\right)^{\frac{2}{3}}. \tag{3.980}$$

The factor B_1 can be numerically approximated using the Python program B1.py (see Appendix A.2). With $B_1 = 0.768$, the critical stress is:

$$\sigma_{\mathrm{cr}} = 27555.434\,\mathrm{MPa} > \sigma_{x,\mathrm{F}}. \tag{3.981}$$

Thus, there is no failure due to local wrinkling of the compression face sheet.

2.5.54 **Failure analysis of a sandwich beam under 3-point bending with point load**

- Internal reactions (Q_z, M_y):

The functions of the internal reactions can be taken from the solution of Exercise 2.2.13:

$$M_y(x) = -\frac{F_0}{2} \times x \quad \text{for} \quad 0 \leq x \leq \frac{L}{2}, \tag{3.982}$$

$$Q_z(x) = -\frac{F_0}{2} \quad \text{for} \quad 0 \leq x \leq \frac{L}{2}, \tag{3.983}$$

where the maximum values are as follows:

$$M_{y,\max} = M_y\left(\tfrac{L}{2}\right) = -\frac{F_0 L}{4}, \tag{3.984}$$

$$Q_{z,\max} = -\frac{F_0}{2}. \tag{3.985}$$

The internal reactions and the corresponding maximum values are shown in Fig. 3.52.

- Theory

Checking whether one of the simplifying theories is applicable:

$$\frac{6E^{\mathrm{F}}\Delta h^{\mathrm{F}}(h_{\mathrm{c}})^2}{E^{\mathrm{C}}(\Delta h^{\mathrm{C}})^3} = 115.625 > 100, \tag{3.986}$$

$$\frac{\Delta h^{\mathrm{C}}}{\Delta h^{\mathrm{F}}} = 4.0 < 4.77. \tag{3.987}$$

Thus, the simplifying theory for soft cores can be applied.

- Maximum normal stress in the face sheets:

Using Eq. $(5.43)_{\mathrm{TB}}$ for soft cores, i.e.

$$\sigma_{x,\mathrm{F}} = \frac{M_{y,\max}}{\dfrac{b(\Delta h^{\mathrm{F}})^3}{6} + \dfrac{b\Delta h^{\mathrm{F}}(h_{\mathrm{c}})^2}{2}} \times \left(\Delta h^{\mathrm{F}} + \frac{\Delta h^{\mathrm{C}}}{2}\right), \tag{3.988}$$

the maximum normal stress in the face sheets with the given numerical values is:

$$\sigma_{x,\mathrm{F}} = 4.737\,\mathrm{MPa} < R^{\mathrm{F}}_{\mathrm{p}0.2}. \tag{3.989}$$

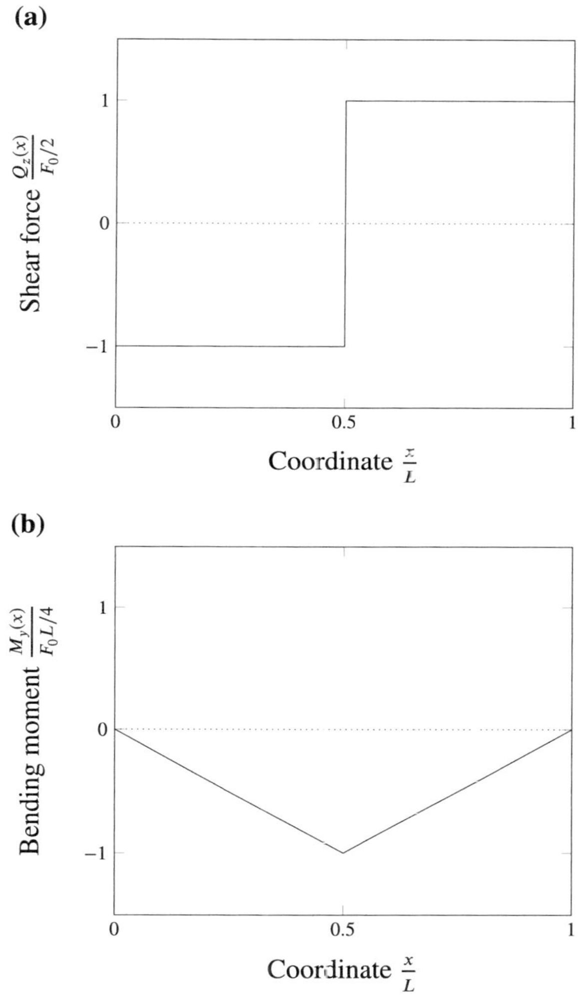

Fig. 3.52 Internal reactions for a sandwich under 3-point bending with point load F_0 in negative z-direction: **a** shear force and **b** bending moment diagram

Thus, there is no failure in the face sheet.

- Maximum shear stress in the core:

Using Eq. $(5.87)_{\text{TB}}$ for soft cores, i.e.

$$\tau_{zx,\text{C}} = \frac{|Q_{z,\max}| h_{\text{c}}}{b\left(\frac{(\Delta h^{\text{F}})^2}{3} + (h_{\text{c}})^2\right)}, \tag{3.990}$$

the maximum shear stress in the core with the given numerical values is:

$$\tau_{zx,\text{C}} = 0.0987\,\text{MPa} < \tau_{\text{aB}}^{\text{C}}. \tag{3.991}$$

Thus, there is no shear failure in the core.

- Shear stress in the adhesive layer:

The shear stress in the adhesive layer is in this case equal to the constant shear stress in the core:

$$\tau_{zx} = 0.0987\,\text{MPa} < \tau_{\text{aB}}. \tag{3.992}$$

Thus, there is no failure in the adhesive layer.

- Local wrinkling of the compression face sheet:

The critical stress is generally calculated according to Eq. $(6.28)_{\text{TB}}$ as:

$$\sigma_{\text{cr}} = B_1 \left(E^{\text{F}}\right)^{\frac{1}{3}} \left(E^{\text{C}}\right)^{\frac{2}{3}}. \tag{3.993}$$

The factor B_1 can be numerically approximated using the Python program B1.py (see Appendix A.2). With $B_1 = 0.648$, the critical stress is:

$$\sigma_{\text{cr}} = 3564.461\,\text{MPa} > \sigma_{x,\text{F}}. \tag{3.994}$$

Thus, there is no failure due to local wrinkling of the compression face sheet.

2.5.55 **Failure analysis of a sandwich beam under 3-point bending with distributed load**

- Internal reactions (Q_z, M_y):

To determine the internal reactions, the inside must be cut free, see Fig. 3.53.

In the Section 1 with $0 \leq x \leq \frac{3L}{8}$ we get:

$$\overset{\curvearrowright}{\sum} M_y = 0: \quad +M_y(x) + \frac{q_0 L}{8} x = 0 \quad \rightarrow \quad M_y(x) = -\frac{q_0 L}{8} x, \tag{3.995}$$

correspondingly for the vertical force equilibrium:

$$\overset{\uparrow}{\sum} F_z = 0: \quad +\frac{q_0 L}{8} + Q_z(x) = 0 \quad \rightarrow \quad Q_z(x) = -\frac{q_0 L}{8}. \tag{3.996}$$

In Section 2 with $\frac{3L}{8} \leq x \leq \frac{5L}{8}$ we get:

$$\overset{\curvearrowright}{\sum} M_y = 0: \quad +\frac{q_0 L}{8} x - \frac{q_0 \left(x - \frac{3L}{8}\right)^2}{2} + M_y(x) = 0 \tag{3.997}$$

$$\rightarrow \quad M_y(x) = \frac{q_0}{2} \left(x - \frac{3L}{8}\right)^2 - \frac{q_0 L}{8} x, \tag{3.998}$$

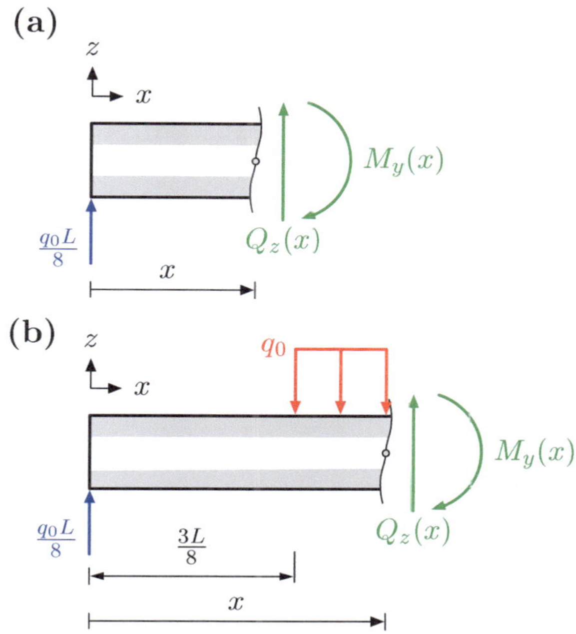

Fig. 3.53 Cutting free inside the beam under 3-point bending with distributed load: **a** Section 11 with $0 \leq x \leq \frac{3L}{8}$ and **b** Section 2 with $\frac{3L}{8} \leq x \leq \frac{5L}{8}$

correspondingly for the vertical force equilibrium:

$$\sum^{\uparrow} F_z = 0: \quad +\frac{q_0 L}{8} + Q_z(x) - q_0\left(x - \frac{3L}{8}\right) = 0 \tag{3.999}$$

$$\rightarrow \quad Q_z(x) = q_0\left(x - \frac{L}{2}\right). \tag{3.1000}$$

The internal reactions and the corresponding maximum values are shown in Fig. 3.54.

- Theory

Checking whether one of the simplifying theories is applicable:

$$\frac{6E^{\mathrm{F}}\Delta h^{\mathrm{F}}(h_{\mathrm{c}})^2}{E^{\mathrm{C}}(\Delta h^{\mathrm{C}})^3} = 115.625 > 100, \tag{3.1001}$$

$$\frac{\Delta h^{\mathrm{C}}}{\Delta h^{\mathrm{F}}} = 4.0 < 4.77. \tag{3.1002}$$

Thus, the simplifying theory for soft cores can be applied.

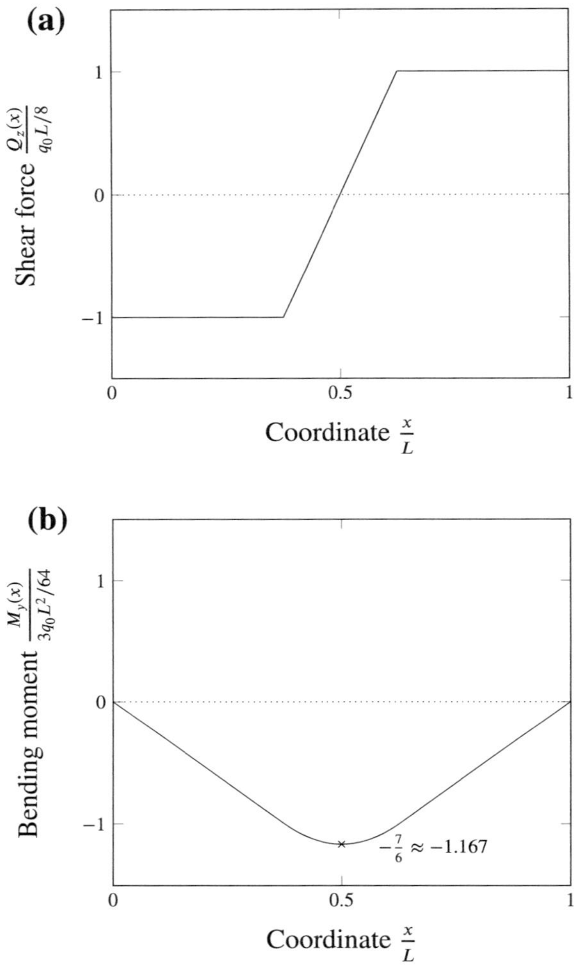

Fig. 3.54 Internal reactions for a sandwich beam under 3-point bending with negative distributed load q_0 in the range $3L/8 \leq x \leq 5L/8$: **a** shear force and **b** bending moment diagram

- Maximum normal stress in the face sheets:

Using Eq. $(5.43)_{\text{TB}}$ for soft cores, i.e.

$$\sigma_{x,\text{F}} = \frac{M_{y,\max}}{\dfrac{b(\Delta h^{\text{F}})^3}{6} + \dfrac{b\Delta h^{\text{F}}(h_{\text{c}})^2}{2}} \times \left(\Delta h^{\text{F}} + \frac{\Delta h^{\text{C}}}{2}\right), \tag{3.1003}$$

the maximum normal stress in the face sheets with the given numerical value is:

$$\sigma_{x,\text{F}} = 4.145\,\text{MPa} < R^{\text{F}}_{\text{p0.2}}. \tag{3.1004}$$

Thus, there is no failure in the face sheet.

• Maximum shear stress in the core:

Using Eq. $(5.87)_{\text{TB}}$ for soft cores, i.e.

$$\tau_{zx,\text{C}} = \frac{|Q_{z,\max}|h_{\text{c}}}{b\left(\frac{(\Delta h^{\text{F}})^2}{3} + (h_{\text{c}})^2\right)}, \tag{3.1005}$$

the maximum shear stress in the core with the given numerical values is:

$$\tau_{zx,\text{C}} = 0.0987\,\text{MPa} < \tau_{\text{aB}}^{\text{C}}. \tag{3.1006}$$

Thus, there is no shear failure in the core.

• Shear stress in the adhesive layer:

The shear stress in the adhesive layer is in this case equal to the constant shear stress in the core:

$$\tau_{zx} = 0.0987\,\text{MPa} < \tau_{\text{aB}}. \tag{3.1007}$$

Thus, there is no failure in the adhesive layer.

• Local wrinkling of the compression face sheet:

The critical stress is generally calculated according to Eq. $(5.28)_{\text{TB}}$ as:

$$\sigma_{\text{cr}} = B_1 \left(E^{\text{F}}\right)^{\frac{1}{3}} \left(E^{\text{C}}\right)^{\frac{2}{3}}. \tag{3.1008}$$

The factor B_1 can be numerically approximated using the Python program B1.py (see Appendix A.2). With $B_1 = 0.648$, the critical stress is:

$$\sigma_{\text{cr}} = 3564.461\,\text{MPa} > \sigma_{x,\text{F}}. \tag{3.1009}$$

Thus, there is no failure due to local wrinkling of the compression face sheet.

2.5.56 Failure analysis of a sandwich beam under 5-point bending with point loads

• Internal reactions (Q_z, M_y):

To determine the internal reactions, the free-body diagram shown in Fig. 3.55a can be used. This results in four sections for determining the internal reactions. Alternatively, taking symmetry into account, the equivalent system shown in Fig. 3.55b can be used.

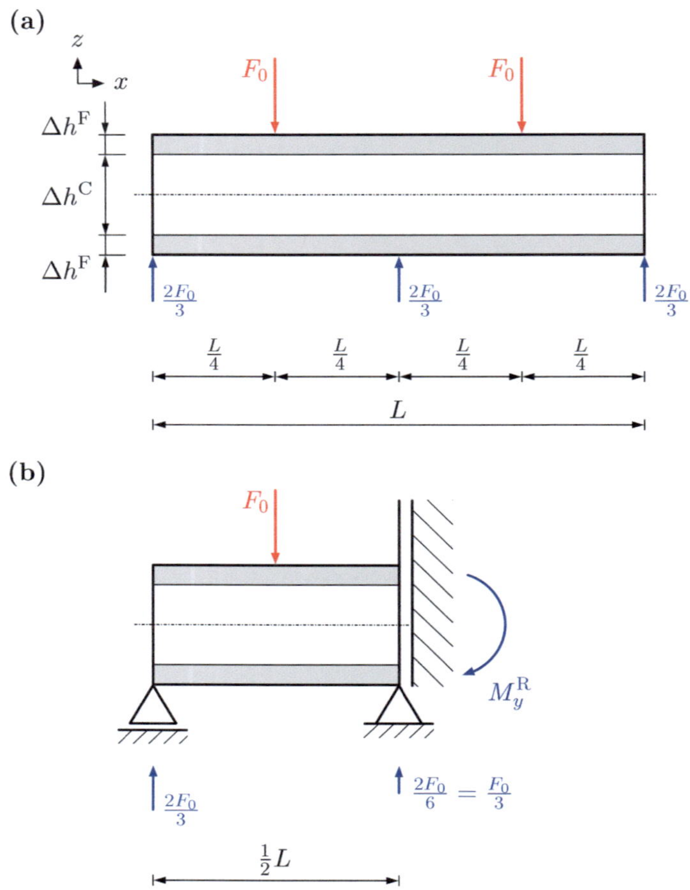

Fig. 3.55 Sandwich beam under 5-point bending with concentrated loads: **a** free body diagram; **b** free body diagram considering symmetry

The internal reactions and the corresponding maximum values are shown in Fig. 3.56.

- Theory

Checking whether one of the simplifying theories is applicable:

$$\frac{6E^{\mathrm{F}}\Delta h^{\mathrm{F}}(h_{\mathrm{c}})^2}{E^{\mathrm{C}}(\Delta h^{\mathrm{C}})^3} = 115.625 > 100, \tag{3.1010}$$

$$\frac{\Delta h^{\mathrm{C}}}{\Delta h^{\mathrm{F}}} = 4.0 < 4.77. \tag{3.1011}$$

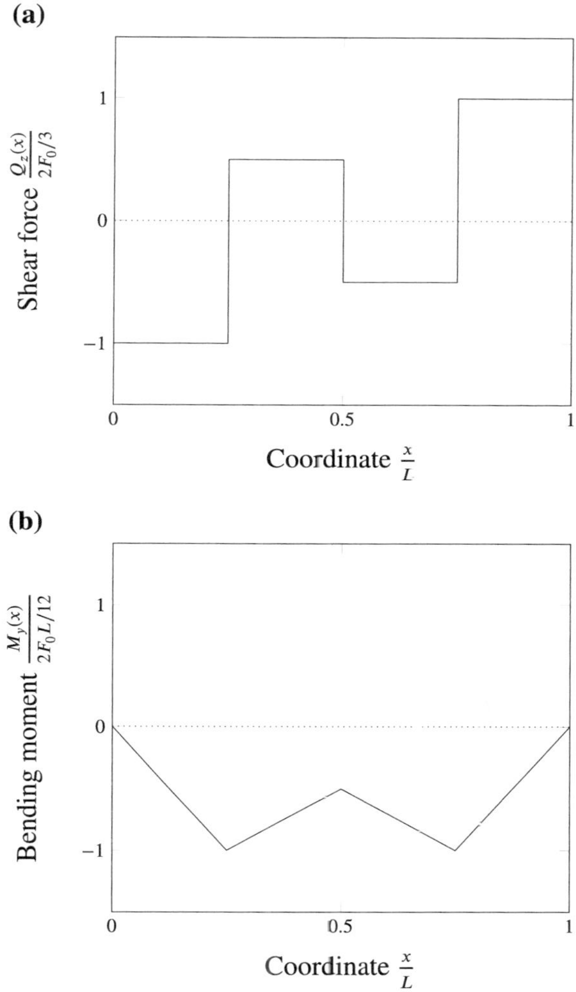

Fig. 3.56 Internal reactions for a sandwich beam under 5-point bending with concentrated loads: **a** shear force and **b** bending moment diagram

Thus, the simplifying theory for soft cores can be applied.

- Maximum normal stress in the face sheets:

Using Eq. $(5.43)_{\text{TB}}$ for soft cores, i.e.

$$\sigma_{x,\text{F}} = \frac{M_{y,\max}}{\dfrac{b(\Delta h^{\text{F}})^3}{6} + \dfrac{b\Delta h^{\text{F}}(h_{\text{c}})^2}{2}} \times \left(\Delta h^{\text{F}} + \frac{\Delta h^{\text{C}}}{2}\right), \tag{3.1012}$$

the maximum normal stress in the face sheets with the given numerical values is:

$$\sigma_{x,\text{F}} = 1.579\,\text{MPa} < R^{\text{F}}_{\text{p0.2}}. \tag{3.1013}$$

Thus, there is no failure in the face sheet.

- Maximum shear stress in the core:

Using Eq. $(5.87)_{\text{TB}}$ for soft cores, i.e.

$$\tau_{zx,\text{C}} = \frac{|Q_{z,\text{max}}|h_{\text{c}}}{b\left(\frac{(\Delta h^{\text{F}})^2}{3} + (h_{\text{c}})^2\right)}, \tag{3.1014}$$

the maximum shear stress in the core with the given numerical values is:

$$\tau_{zx,\text{C}} = 0.0658\,\text{MPa} < \tau_{\text{aB}}^{\text{C}}. \tag{3.1015}$$

Thus, there is no shear failure in the core.

- Shear stress in the adhesive layer:

The shear stress in the adhesive layer is in this case equal to the constant shear stress in the core:

$$\tau_{zx} = 0.0658\,\text{MPa} < \tau_{\text{aB}}. \tag{3.1016}$$

Thus, there is no failure in the adhesive layer.

- Local wrinkling of the compression face sheet:

The critical stress is generally calculated according to Eq. $(6.28)_{\text{TB}}$ as:

$$\sigma_{\text{cr}} = B_1 \left(E^{\text{F}}\right)^{\frac{1}{3}} \left(E^{\text{C}}\right)^{\frac{2}{3}}. \tag{3.1017}$$

The factor B_1 can be numerically approximated using the Python program B1.py (see Appendix A.2). With $B_1 = 0.648$, the critical stress is:

$$\sigma_{\text{cr}} = 3564.461\,\text{MPa} > \sigma_{x,\text{F}}. \tag{3.1018}$$

Thus, there is no failure due to local wrinkling of the compression face sheet.

2.5.57 **Failure analysis of a sandwich beam under 5-point bending with distributed loads**

- Internal reactions (Q_z, M_y):

To determine the internal reactions, the free-body diagram shown in Fig. 3.57a can be used. This results in five sections for determining the internal reactions. Alternatively, taking symmetry into account, the equivalent system shown in Fig. 3.57b can be used.

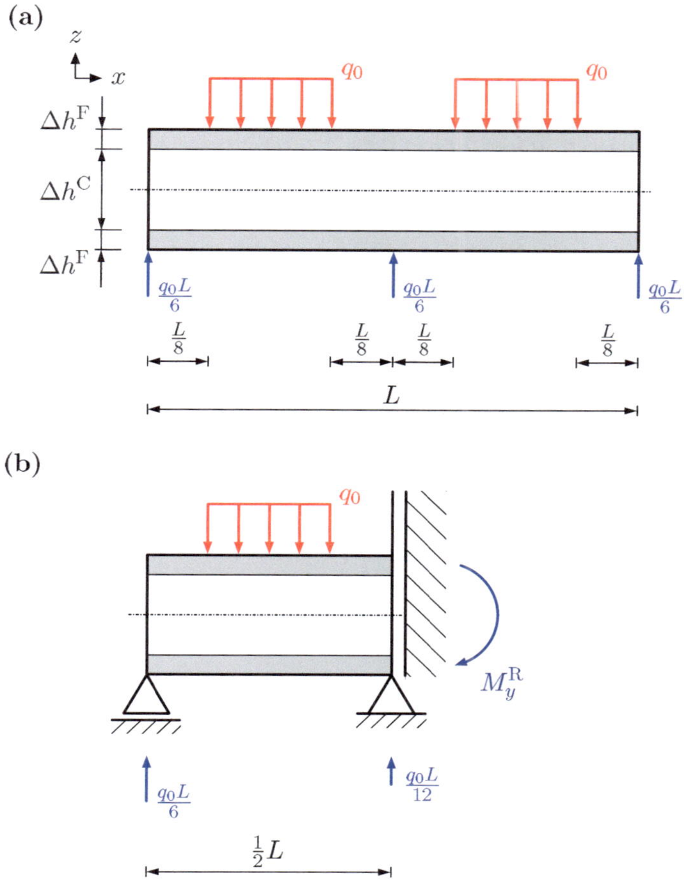

Fig. 3.57 Sandwich beam under 5-point bending with distributed loads: **a** free body diagram; **b** free body diagram considering symmetry

The internal reactions and the corresponding maximum values are shown in Fig. 3.58.

• Theory

Checking whether one of the simplifying theories is applicable:

$$\frac{6E^F \Delta h^F (h_c)^2}{E^C (\Delta h^C)^3} = 115.625 > 100, \tag{3.1019}$$

$$\frac{\Delta h^C}{\Delta h^F} = 4.0 < 4.77. \tag{3.1020}$$

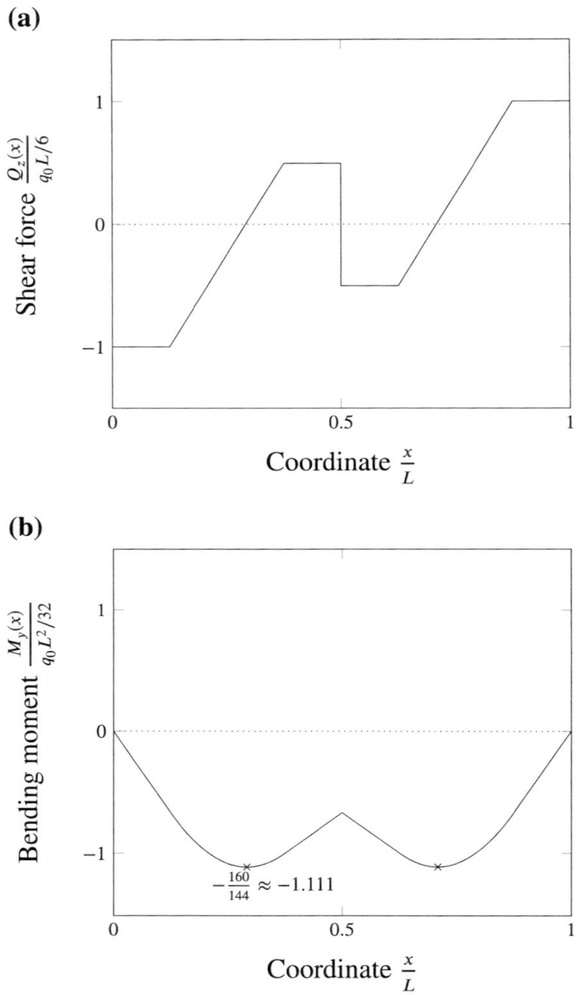

Fig. 3.58 Internal reactions for a sandwich beam under 5-point bending with distributed load: **a** shear force and **b** bending moment diagram

Thus, the simplifying theory for soft cores can be applied.

- Maximum normal stress in the face sheets:

Using Eq. (5.43)$_{\text{TB}}$ for soft cores, i.e.

$$\sigma_{x,\text{F}} = \frac{M_{y,\text{max}}}{\frac{b(\Delta h^{\text{F}})^3}{6} + \frac{b\Delta h^{\text{F}}(h_{\text{c}})^2}{2}} \times \left(\Delta h^{\text{F}} + \frac{\Delta h^{\text{C}}}{2}\right), \tag{3.1021}$$

the maximum normal stress in the face sheets with the given numerical values is:

$$\sigma_{x,\text{F}} = 0.177\,\text{MPa} < R^{\text{F}}_{\text{p0.2}}. \tag{3.1022}$$

Thus, there is no failure in the face sheet.

- Maximum shear stress in the core:

Using Eq. (5.87)TB for soft cores, i.e.

$$\tau_{zx,\mathrm{C}} = \frac{|Q_{z,\max}|h_\mathrm{c}}{b\left(\frac{(\Delta h^\mathrm{F})^2}{3} + (h_\mathrm{c})^2\right)}, \tag{3.1023}$$

the maximum shear stress in the core with the given numerical values is:

$$\tau_{zx,\mathrm{C}} = 0.0658\,\mathrm{MPa} < \tau_{\mathrm{aB}}^{\mathrm{C}}. \tag{3.1024}$$

Thus, there is no shear failure in the core.

- Shear stress in the adhesive layer:

The shear stress in the adhesive layer is in this case equal to the constant shear stress in the core:

$$\tau_{zx} = 0.0658\,\mathrm{MPa} < \tau_{\mathrm{aB}}. \tag{3.1025}$$

Thus, there is no failure in the adhesive layer.

- Local wrinkling of the compression face sheet:

The critical stress is generally calculated according to Eq. (6.28)TB as:

$$\sigma_\mathrm{cr} = B_1 \left(E^\mathrm{F}\right)^{\frac{1}{3}} \left(E^\mathrm{C}\right)^{\frac{2}{3}}. \tag{3.1026}$$

The factor B_1 can be numerically approximated using the Python program B1.py (see Appendix A.2). With $B_1 = 0.648$, the critical stress is:

$$\sigma_\mathrm{cr} = 3564.461\,\mathrm{MPa} > \sigma_{x,\mathrm{F}}. \tag{3.1027}$$

Thus, there is no failure due to local wrinkling of the compression face sheet.

2.5.58 **Global instability failure of a sandwich beam clamped on both sides under compressive loading**

The configuration for determining the internal reactions is shown in Fig. 3.59.

This results in the following internal reactions:

$$M_y(x) = F u_z(x) - M_y^\mathrm{R}, \tag{3.1028}$$

$$Q_z(x) = F\frac{\mathrm{d}u_z(x)}{\mathrm{d}x}. \tag{3.1029}$$

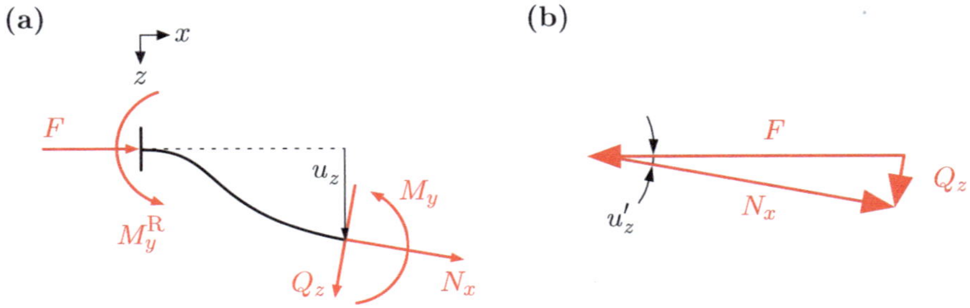

Fig. 3.59 Sandwich beam clamped on both sides under compressive load: **a** cutting free at location x; **b** force triangle (without considering the current sign)

The describing differential equation results in:

$$\frac{\mathrm{d}^2 u_z(x)}{\mathrm{d}x^2} + \lambda^2 u_z(x) = \lambda^2 \frac{M_y^{\mathrm{R}}}{F} \quad \text{with} \quad \lambda^2 = \frac{F}{\overline{EI_y}\left(1 - \frac{1}{AG^{\mathrm{C}}}\right)}. \tag{3.1030}$$

From the general solution,[3] i.e. $u_z(x) = c_1 \cos(\lambda x) + c_2 \sin(\lambda x) + \frac{M_y^{\mathrm{R}}}{F}$, results with the corresponding boundary conditions, i.e. $u_z(0) = 0$ and $\frac{\mathrm{d}u_z(0)}{\mathrm{d}x} = 0$, the integration constants to $c_1 = -\frac{M_y^{\mathrm{R}}}{F}$ and $c_2 = 0$. Finally, the additional boundary condition $u_z(L) = 0$ results in the condition for determining the buckling force: $\cos(\lambda x) = 1$. This results in the buckling force for the boundary conditions under consideration:

$$F_{\mathrm{cr}} = \frac{\frac{4\pi^2 \overline{EI_y}}{L^2}}{1 + \frac{1}{AG^{\mathrm{C}} \times \frac{4\pi^2 \overline{EI_y}}{L^2}}} = \frac{F_{\mathrm{cr}}^{\mathrm{E}}}{1 + \frac{F_{\mathrm{cr}}^{\mathrm{E}}}{AG^{\mathrm{C}}}}. \tag{3.1031}$$

2.5.59 **Instability failure of a sandwich beam hinged on both ends under compressive loading**

- Verification of conditions for soft core and thin face sheets:

$$\frac{6E^{\mathrm{F}}\Delta h^{\mathrm{F}}(h_{\mathrm{c}})^2}{E^{\mathrm{C}}(\Delta h^{\mathrm{C}})^3} = 122.38 \geq 100, \tag{3.1032}$$

$$\frac{\Delta h^{\mathrm{C}}}{\Delta h^{\mathrm{F}}} = 20.0 \geq 4.77. \tag{3.1033}$$

Thus, the conditions for a soft core and thin face sheets are met.

[3] This can be determined with a computer algebra system (e.g. Maxima).

• Global buckling:

Using Eq. $(6.19)_{\mathrm{TB}}$ results in:

$$F_{\mathrm{cr}} = \frac{\frac{\pi^2 \overline{EI_y}}{L^2}}{1 + \frac{1}{AG^{\mathrm{C}} \times \frac{\pi^2 \overline{EI_y}}{L^2}}} = \frac{F_{\mathrm{cr}}^{\mathrm{E}}}{1 + \frac{F_{\mathrm{cr}}^{\mathrm{E}}}{AG^{\mathrm{C}}}} = 614081.55\ \mathrm{N}, \tag{3.1034}$$

$$\sigma_{\mathrm{cr}} = \frac{F_{\mathrm{cr}}}{b(\Delta h^{\mathrm{C}} + 2\Delta h^{\mathrm{F}})} = 27.91\ \mathrm{MPa}. \tag{3.1035}$$

• Local wrinkling, antisymmetric:

The critical stress is generally calculated according to Eq. $(6.39)_{\mathrm{TB}}$ as:

$$\sigma_{\mathrm{cr}} = B_1 \left(E^{\mathrm{F}}\right)^{\frac{1}{3}} \left(E^{\mathrm{C}}\right)^{\frac{2}{3}}. \tag{3.1036}$$

The factor B_1 can be numerically approximated using the Python program B1.py (see Appendix A.2). With $k = 0.359$, this results in $B_1 = 0.492$ and $\Theta = 2.362$. Thus, the critical stress is:

$$\sigma_{\mathrm{cr}} = 706.236\,\mathrm{MPa}. \tag{3.1037}$$

• Local wrinkling, symmetric:

The critical stress is generally calculated according to Eq. $(6.43)_{\mathrm{TB}}$ as:

$$\sigma_{\mathrm{cr}} = B_1 \left(E^{\mathrm{F}}\right)^{\frac{1}{3}} \left(E^{\mathrm{C}}\right)^{\frac{2}{3}}. \tag{3.1038}$$

The factor B_1 can be numerically approximated using the Python program B1.py (see Appendix A.2). With $k = 0.359$, this results in $B_1 = 0.651$ and $\Theta = 4.754$. Thus, the critical stress is:

$$\sigma_{\mathrm{cr}} = 935.132\,\mathrm{MPa}. \tag{3.1039}$$

• Course of the critical buckling stress (Fig. 3.60)
The limit of the buckling stress for $L \to 0$ is:

$$\lim_{L \to 0} \sigma_{\mathrm{cr}} = \frac{1}{A} \times \frac{\frac{1}{L^2}}{\frac{1}{L^2}} \times \frac{\pi^2 \overline{EI_y}}{L^2 + \frac{\pi^2 \overline{EI_y}}{AG^{\mathrm{C}}}} \to G^{\mathrm{C}}. \tag{3.1040}$$

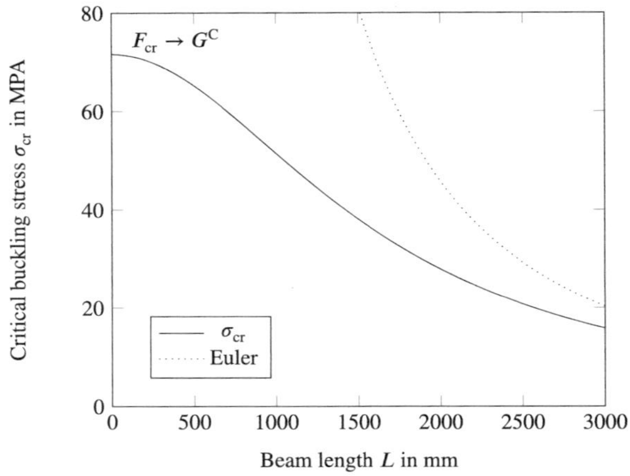

Fig. 3.60 Course of the critical buckling stress as a function of the beam length

3.6 Sandwich Structures: Optimization

The Python programs Optm_Comp.py and Optm_Comp_Bend_1.py from the Appendix A.2 can be used to solve the following optimization problems.

2.6.60 **Optimization of a sandwich beam under compressive loading**

• Conversion of densities to consistent units

Since the stiffnesses are given in MPa = N/mm^2, the densities must be converted to the consistent unit N/mm^3. The result is the following:

$$\varrho^{\mathrm{F}} = 2691\,\mathrm{kg/m^3} = 2691 \times 9.81 \times 10^{-9}\,\mathrm{N/mm^3}, \tag{3.1041}$$

$$\varrho^{\mathrm{C}} = 240\,\mathrm{kg/m^3} = 240 \times 9.81 \times 10^{-9}\,\mathrm{N/mm^3}. \tag{3.1042}$$

Case (a): $F = 2670\,\mathrm{N}$:

• Calculation of point E:

$$E\left(\Delta h_E^{\mathrm{C,n}};\, \Delta h_E^{\mathrm{F,n}}\right) = \left(0.0191215;\, 2.923250 \times 10^{-5}\right). \tag{3.1043}$$

• Calculation of point G:

$$G\left(\Delta h_G^{\mathrm{C,n}};\, \Delta h_G^{\mathrm{F,n}}\right) = \left(0.00820895;\, 1.711632 \times 10^{-4}\right). \tag{3.1044}$$

• Optimal design:

Since $\Delta h_G^{\mathrm{C,n}} \leq \Delta h_E^{\mathrm{C,n}}$ holds, point G results as the optimal geometry with:

$$\Delta h_G^{\mathrm{C}} = \Delta h_G^{\mathrm{C,n}} \times L = 20.85\,\mathrm{mm} \tag{3.1045}$$

$$\Delta h_G^{\mathrm{F}} = \Delta h_G^{\mathrm{F,n}} \times L = 0.43\,\mathrm{mm}. \tag{3.1046}$$

Under certain circumstances, however, one still has to respect the minimum thicknesses of sheet metal.

Case (b): $10 \times F = 26700\,\mathrm{N}$:

• Calculation of point E:

$$E\left(\Delta h_E^{\mathrm{C,n}}; \Delta h_E^{\mathrm{F,n}}\right) = \left(0.0242730; 2.923250 \times 10^{-4}\right). \tag{3.1047}$$

• Calculation of point G

$$G\left(\Delta h_G^{\mathrm{C,n}}; \Delta h_G^{\mathrm{F,n}}\right) = \left(0.0225247; 3.589804 \times 10^{-4}\right). \tag{3.1048}$$

• Optimal design

Since $\Delta h_G^{\mathrm{C,n}} \leq \Delta h_E^{\mathrm{C,n}}$ holds, point G results as the optimal geometry with:

$$\Delta h_G^{\mathrm{C}} = \Delta h_G^{\mathrm{C,n}} \times L = 57.21\,\mathrm{mm}, \tag{3.1049}$$

$$\Delta h_G^{\mathrm{F}} = \Delta h_G^{\mathrm{F,n}} \times L = 0.91\,\mathrm{mm}. \tag{3.1050}$$

Under certain circumstances, however, one still has to respect the minimum thicknesses of sheet metal.

The position of the two minima is shown in Fig. 3.61. It is possible to conclude that the position of the minimum shifts further to the right with increasing external load.

2.6.61 **Optimization of a sandwich beam under bending load due to a single force**

• Second pole of the function g_3 in the $\Delta h^{\mathrm{C,n}}$-$\Delta h^{\mathrm{F,n}}$ coordinate system

$$\Delta h^{\mathrm{C,n}} = \frac{F}{4bLG^{\mathrm{C}}r_1} = 0.0832179. \tag{3.1051}$$

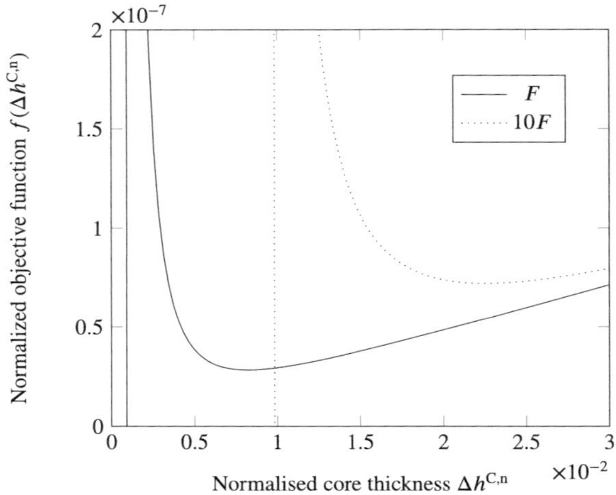

Fig. 3.61 Normalized objective function depending on the normalized core thickness with the position of the minima for different loads

- Intersection points in the $\Delta h^{\text{C,n}}$-$\Delta h^{\text{F,n}}$ coordinate system

Point E (g_1–g_2)

$$E = \left(\Delta h_E^{\text{C,n}} \,\middle|\, \Delta h_E^{\text{F,n}}\right) = \left(\frac{F_0}{2bL\tau_{\text{p}}} \,\middle|\, \frac{\tau_{\text{p}}}{2\sigma_{\text{cr}}}\right) \tag{3.1052}$$

$$= (0.0124827|\, 0.00116961)\,. \tag{3.1053}$$

Point A (g_1–g_3)

$$A = \left(\Delta h_A^{\text{C,n}} \,\middle|\, \Delta h_A^{\text{D,n}}\right) = \left(\frac{4}{Lr_1}\left(\frac{2L\sigma_{\text{cr}}}{48E^{\text{D}}} + \frac{F_0}{16bG^{\text{K}}}\right) \,\middle|\, \frac{F_0}{4bL\sigma_{\text{cr}}\Delta h^{\text{K,n}}}\right) \tag{3.1054}$$

$$= (0.130717|\, 1.116902 \times 10^{-4})\,. \tag{3.1055}$$

- Minima of the objective function f along the constraints

Point B (minimum of f along g_1)

$$\frac{\partial f(\Delta h^{\text{C,n}})}{\partial \Delta h^{\text{C,n}}} \stackrel{!}{=} 0 \quad \Rightarrow \quad \Delta h_B^{\text{C,n}} = \sqrt{\frac{1}{2} \times \frac{\varrho^{\text{F}}}{\varrho^{\text{C}}} \times \frac{F_0}{bL\sigma_{\text{cr}}}}\,. \tag{3.1056}$$

$$B = \left(\Delta h_B^{\text{C,n}} \,\middle|\, \Delta h_B^{\text{F,n}}\right) = \left(\sqrt{\frac{1}{2} \times \frac{\varrho^{\text{F}}}{\varrho^{\text{C}}} \times \frac{F_0}{bL\sigma_{\text{cr}}}} \,\middle|\, \sqrt{\frac{1}{8} \times \frac{\varrho^{\text{C}}}{\varrho^{\text{F}}} \times \frac{F_0}{bL\sigma_{\text{cr}}}}\right) \tag{3.1057}$$

$$= (0.0180942|\, 8.0687762 \times 10^{-4})\,. \tag{3.1058}$$

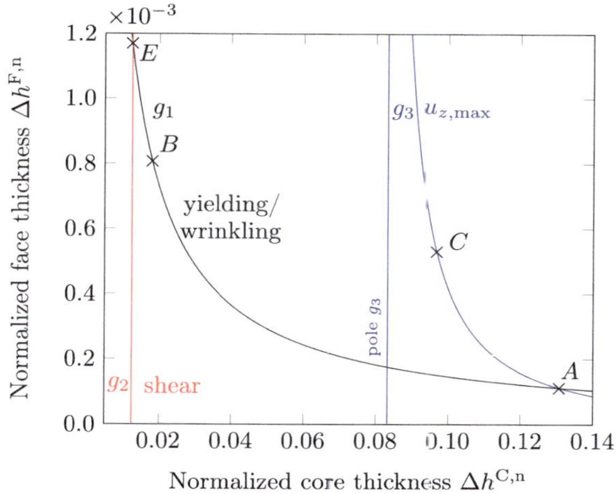

Fig. 3.62 Normalized face thickness as a function of normalized core thickness based on functions g_1, g_2 and g_3 for a bending beam with center load

Point C (minimum of f along g_3)

$$\frac{\partial f(\Delta h^{\mathrm{C,n}})}{\partial \Delta h^{\mathrm{C,n}}} \stackrel{!}{=} 0 \quad \Rightarrow \quad \Delta h_C^{\mathrm{C,n}} \text{ (Newton's iteration).} \tag{3.1059}$$

$$C = \left(\Delta h_C^{\mathrm{C,n}} \;\middle|\; \Delta h_C^{\mathrm{F,n}}\right) \tag{3.1060}$$

$$= (0.0967536 |\, 5.295226 \times 10^{-4}) \,. \tag{3.1061}$$

• Optimum point

It is case 1 and therefore point C is the optimal point.

Optimal dimensions: $\Delta h^{\mathrm{C}} = 245.754$ mm, $\Delta h^{\mathrm{F}} = 1.345$ mm. With these geometric dimensions, the condition for thin face sheets and a soft core is also met:

$$\frac{6E^{\mathrm{F}}\Delta h^{\mathrm{F}}(h_{\mathrm{c}})^2}{E^{\mathrm{C}}(\Delta h^{\mathrm{C}})^3} = 328.4 \geq 100, \tag{3.1062}$$

$$\frac{\Delta h^{\mathrm{C}}}{\Delta h^{\mathrm{F}}} = 182.7 \geq 4.77. \tag{3.1063}$$

• Graphical representation in the $\Delta h^{\mathrm{C,n}}$-$\Delta h^{\mathrm{F,n}}$ coordinate system (see Fig. 3.62).

2.6.62 **Optimization of a sandwich beam under bending load due to a distributed load**

• Objective function and constraints

$$f\left(\Delta h^{\mathrm{F,n}}, \Delta h^{\mathrm{C,n}}\right) = \varrho^{\mathrm{C}} \Delta h^{\mathrm{C,n}} + 2\varrho^{\mathrm{F}} \Delta h^{\mathrm{F,n}}. \tag{3.1064}$$

$$g_1(\Delta h^{\mathrm{C,n}}, \Delta h^{\mathrm{F,n}}) = \frac{q_0}{8\Delta h^{\mathrm{C,n}}\Delta h^{\mathrm{F,n}}b} < \sigma_{\mathrm{cr}}, \tag{3.1065}$$

$$g_2(\Delta h^{\mathrm{C,n}}, \Delta h^{\mathrm{F,n}}) = \frac{q_0}{2\Delta h^{\mathrm{C,n}}b} < \tau_{\mathrm{p}}, \tag{3.1066}$$

$$g_3(\Delta h^{\mathrm{C,n}}, \Delta h^{\mathrm{F,n}}) = \frac{10q_0}{384E^{\mathrm{F}}b\Delta h^{\mathrm{F,n}}(\Delta h^{\mathrm{C,n}})^2} + \frac{q_0}{8bG^{\mathrm{C}}\Delta h^{\mathrm{C,n}}} < r_1. \tag{3.1067}$$

Representation of the constraints in the $\Delta h^{\mathrm{C,n}}$-$\Delta h^{\mathrm{F,n}}$ coordinate system:

$$g_1: \qquad \Delta h^{\mathrm{F,n}}_{g_1} > \frac{q_0}{8b\sigma_{\mathrm{cr}}\Delta h^{\mathrm{C,n}}}, \tag{3.1068}$$

$$g_2: \qquad \Delta h^{\mathrm{C,n}}_{g_2} > \frac{q_0}{2b\tau_{\mathrm{p}}}, \tag{3.1069}$$

$$g_3: \qquad \Delta h^{\mathrm{F,n}}_{g_3} > \frac{\frac{10q_0}{384E^{\mathrm{F}}b(\Delta h^{\mathrm{C,n}})^2}}{r_1 - \frac{q_0}{8bG^{\mathrm{C}}\Delta h^{\mathrm{C,n}}}}. \tag{3.1070}$$

- Intersection points in the $\Delta h^{\mathrm{C,n}}$-$\Delta h^{\mathrm{F,n}}$ coordinate system

Point E (g_1–g_2)

$$E = \left(\Delta h^{\mathrm{C,n}}_E \,\middle|\, \Delta h^{\mathrm{F,n}}_E\right) = \left(\frac{q_0}{2b\tau_{\mathrm{p}}}\,\middle|\, \frac{\tau_{\mathrm{p}}}{4\sigma_{\mathrm{cr}}}\right) \tag{3.1071}$$

$$= \left(0.0124827 \middle|\; 5.848035 \times 10^{-4}\right). \tag{3.1072}$$

Point A (g_1–g_3)

$$A = \left(\Delta h^{\mathrm{C,n}}_A \,\middle|\, \Delta h^{\mathrm{F,n}}_A\right) = \left(\frac{1}{r_1 b}\left(\frac{80b\sigma_{\mathrm{cr}}}{384E^{\mathrm{F}}} + \frac{q_0}{8G^{\mathrm{C}}}\right)\,\middle|\, \frac{q_0}{8b\sigma_{\mathrm{cr}}\Delta h^{\mathrm{C,n}}}\right) \tag{3.1073}$$

$$= \left(0.100983 \middle|\; 7.228848 \times 10^{-5}\right). \tag{3.1074}$$

- Minima of the objective function f along the constraints

Point B (minimum of f along g_1)

$$\frac{\partial f(\Delta h^{\mathrm{C,n}})}{\partial \Delta h^{\mathrm{C,n}}} \overset{!}{=} 0 \;\Rightarrow\; \Delta h^{\mathrm{C,n}}_B = \sqrt{\frac{2\varrho^{\mathrm{F}}q_0}{8b\sigma_{\mathrm{cr}}\varrho^{\mathrm{C}}}}. \tag{3.1075}$$

$$B = \left(\Delta h_B^{\mathrm{C,n}} \,\middle|\, \Delta h_B^{\mathrm{F,n}}\right) = \left(\sqrt{\frac{2\varrho^{\mathrm{F}} q_0}{8b\sigma_{\mathrm{cr}}\varrho^{\mathrm{C}}}} \,\middle|\, \frac{q_0}{8b\sigma_{\mathrm{cr}}\Delta h_B^{\mathrm{C,n}}}\right) \tag{3.1076}$$

$$= (0.0127946|\ 5.705486 \times 10^{-4})\,. \tag{3.1077}$$

Point C (minimum of f along g_3)

$$\frac{\partial f(\Delta h^{\mathrm{C,n}})}{\partial \Delta h^{\mathrm{C,n}}} \overset{!}{=} 0 \;\Rightarrow\; \Delta h_C^{\mathrm{C,n}} \text{ (Newton's iteration)}. \tag{3.1078}$$

$$C = \left(\Delta h_C^{\mathrm{C,n}} \,\middle|\, \Delta h_C^{\mathrm{F,n}}\right) \tag{3.1079}$$

$$= (0.0563599|\ 5.213426 \times 10^{-4})\,. \tag{3.1080}$$

- Optimum point

$$\Delta h_B^{\mathrm{C,n}} > \Delta h_E^{\mathrm{C,n}} \text{ but } \Delta h_B^{\mathrm{C,n}} < \Delta h_A^{\mathrm{C,n}} \;\Rightarrow\; \text{not a valid point,} \tag{3.1081}$$

$$\Delta h_D^{\mathrm{C,n}} < \Delta h_C^{\mathrm{C,n}} < \Delta h_A^{\mathrm{C,n}} \;\Rightarrow\; \text{optimum point.} \tag{3.1082}$$

Optimal dimensions: $\Delta h^{\mathrm{C}} = 143.154$ mm, $\Delta h^{\mathrm{F}} = 1.324$ mm. With these geometric dimensions, the condition for thin face sheets and a soft core is also met:

$$\frac{6E^{\mathrm{F}}\Delta h^{\mathrm{F}}(h_{\mathrm{c}})^2}{E^{\mathrm{C}}(\Delta h^{\mathrm{C}})^3} = 555.0 \geq 100, \tag{3.1083}$$

$$\frac{\Delta h^{\mathrm{C}}}{\Delta h^{\mathrm{F}}} = 108.1 \geq 4.77. \tag{3.1084}$$

- Graphical representation in the $\Delta h^{\mathrm{C,n}}$-$\Delta h^{\mathrm{F,n}}$ coordinate system (see Fig. 3.63).

2.6.63 **Optimization of a homogeneous beam under bending load due to a single force**

- Shear force and bending moment distribution

The absolute values of the maxima can be taken from Fig. 3.52 as:

$$|Q_{z,\max}| = \frac{F_0}{2} \quad \text{and} \quad |M_{y,\max}| = \frac{F_0 L}{4}. \tag{3.1085}$$

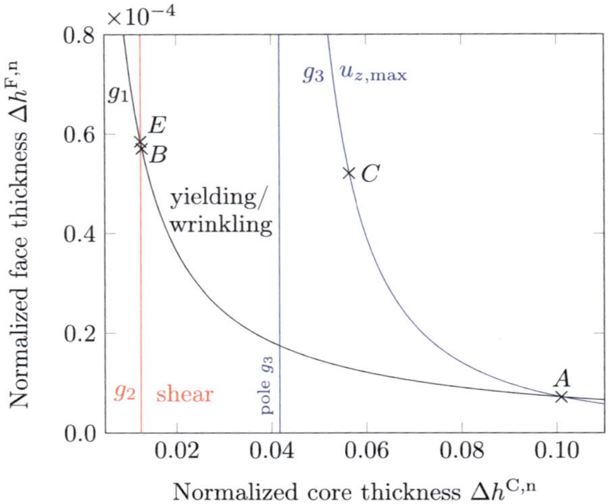

Fig. 3.63 Normalized face thickness as a function of normalized core thickness based on functions g_1, g_2 and g_3 for a bending beam with constant distributed load

- Formulation of the mass and constraints

$$\sigma_{x,\max} = \frac{M_{y,\max}}{I_y} \times \frac{h}{2} < R_{\mathrm{p}0.2}, \tag{3.1086}$$

$$\tau_{xz,\max} = \frac{Q_{z,\max} h^2}{8 I_y} < \tau_{\mathrm{p}}, \tag{3.1087}$$

$$u_{z,\max} = \frac{F_0 L^3}{48 E I_y} < r_1 L, \tag{3.1088}$$

$$h \leq 20 b, \tag{3.1089}$$

$$m = \varrho V. \tag{3.1090}$$

- Normalized representation of mass and constraints

The following formulation results from the normalization $b^{\mathrm{n}} = b/L$ and $h^{\mathrm{n}} = h/L$:

$$g_1(h^{\mathrm{n}}, b^{\mathrm{n}}) = \frac{3 F_0}{2 b^{\mathrm{n}} (h^{\mathrm{n}})^3 L^2} < R_{\mathrm{p}0.2}, \tag{3.1091}$$

$$g_2(h^{\mathrm{n}}, b^{\mathrm{n}}) = \frac{3 F_0}{4 b^{\mathrm{n}} h^{\mathrm{n}} L^2} < \tau_{\mathrm{p}}, \tag{3.1092}$$

$$g_3(h^{\mathrm{n}}, b^{\mathrm{n}}) = \frac{F_0}{4 E b^{\mathrm{n}} (h^{\mathrm{n}})^3 L} < r_1 L, \tag{3.1093}$$

$$g_4(h^{\mathrm{n}}, b^{\mathrm{n}}) = h^{\mathrm{n}} \leq 20 b^{\mathrm{n}}, \tag{3.1094}$$

$$f(h^{\mathrm{n}}, b^{\mathrm{n}}) = \frac{m}{L^3} = \varrho b^{\mathrm{n}} h^{\mathrm{n}}. \tag{3.1095}$$

Representation in the h^n-b^n coordinate system, i.e. rearranged for b^n:

$$g_1: \qquad b^n_{g_1} > \frac{3F_0}{2L^2 R_{p0.2}(h^n)^2}, \tag{3.1096}$$

$$g_2: \qquad b^n_{g_2} > \frac{3F_0}{4L^2 \tau_p h^n}, \tag{3.1097}$$

$$g_3: \qquad b^n_{g_3} > \frac{F_0}{4Er_1 L^2 (h^n)^3}, \tag{3.1098}$$

$$g_4: \qquad b^n_{g_4} \geq \frac{h^n}{20}. \tag{3.1099}$$

Figure 3.64 shows the four constraints g_1, g_2, g_3 and g_4 in the h^n-b^n plane.

• Intersections:

Point A (g_1–g_3)

$$A = \left(h^n_A \mid b^n_A\right) = \left(\frac{R_{p0.2}}{6Er_1} \middle| \frac{54F_0 E^2 r_1^2}{L^2 (R_{p0.2})^3}\right). \tag{3.1100}$$

Point B (g_1–g_4)

$$B = \left(h^n_B \mid b^n_B\right) = \left(\left(\frac{30F_0}{L^2 R_{p0.2}}\right)^{\frac{1}{3}} \middle| \frac{1}{20}\left(\frac{30F_0}{L^2 R_{p0.2}}\right)^{\frac{1}{3}}\right). \tag{3.1101}$$

Point C (g_3–g_4)

$$C = \left(h^n_C \mid b^n_C\right) = \left(\left(\frac{5F_0}{Er_1 L^2}\right)^{\frac{1}{4}} \middle| \frac{1}{2}\left(\frac{5F_0}{Er_1 L^2}\right)^{\frac{1}{4}}\right). \tag{3.1102}$$

Note: The objective function f has no local minimum along g_1 and g_3. The larger h^n, the smaller f (strictly decreasing functions). According to Fig. 3.64, $r_1 = 0.03$ results in point B and $r_1 = 0.003$ in point C as the optimal point.

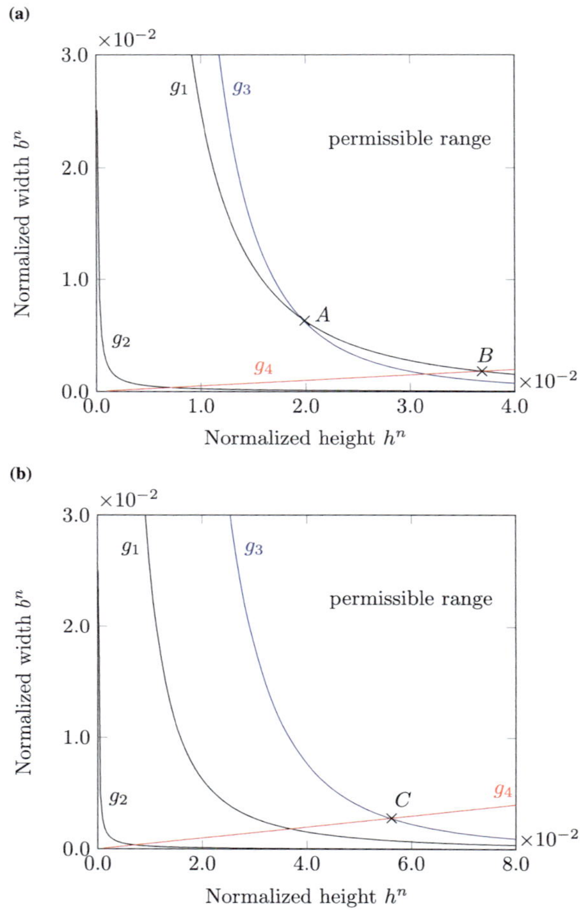

Fig. 3.64 Determination of the optimum point in the h^n-b^n coordinate system: **a** point B as the intersection of g_1 and g_4 ($r_1 = 0.03$). **b** Point C as the intersection of g_3 and g_4 ($r_1 = 0.003$)

References

Javanbakht, Z., Öchsner, A.: Advanced Finite Element Simulation with MSC Marc: Application of User Subroutines. Springer, Cham (2017)

Öchsner, A.: Elasto-Plasticity of Frame Structure Elements: Modeling and Simulation of Rods and Beams. Springer, Berlin (2014)

Öchsner, A.: Computational Statics and Dynamics: An Introduction Based on the Finite Element Method. Springer, Singapore (2016)

Öchsner, A.: A Project-Based Introduction to Computational Statics. Springer, Cham (2018)

Appendix

A.1 Mechanics and Mathematics

A.1.1 Derivation of the Shear Stress Distribution for the Beam

As a starting point for deriving the shear stress distribution for a beam, for example, a cantilever with shear force loading F_0 can be considered, see Fig. A.1. Furthermore, in the following we only consider a rectangular cross section with side dimensions $b \times h$.

An infinitesimal beam element of this configuration is shown in Fig. A.2. The internal reactions shown are drawn according to the common sign definition as indicated in Fig. 2.8$_{\text{TB}}$. Since there is no distributed load, i.e. $q_z = 0$, the vertical force equilibrium gives $Q_z(x) \approx Q_z(x + \mathrm{d}x)$.

The next step is to replace the inner reactions, i.e. the shear force and the bending moment, by the corresponding normal and shear stresses. To do this, a small element of height $h/2 - z'$ is cut out in the vertical direction from the infinitesimal (in the horizontal direction) beam element, see Fig. A.3.

The horizontal forces equilibrium gives for this beam element:

$$-\int \sigma_x(x)\,\underbrace{b(z)\mathrm{d}z}_{\mathrm{d}A} + \int \sigma_x(x + \mathrm{d}x)\,\underbrace{b(z)\mathrm{d}z}_{\mathrm{d}A} - \tau_{zx}(z)b(z)\mathrm{d}x = 0\,, \tag{A.1}$$

or simplified after to a Taylor series expansion of the stress at $(x + \mathrm{d}x)$:

$$\int \frac{\mathrm{d}\sigma_x(x)}{\mathrm{d}x}\mathrm{d}x\mathrm{d}A - \tau_{zx}(z)b(z)\mathrm{d}x = 0\,, \tag{A.2}$$

A. Öchsner, *Exercises and Solutions in Lightweight Design*,
https://doi.org/10.1007/978-3-658-49333-2

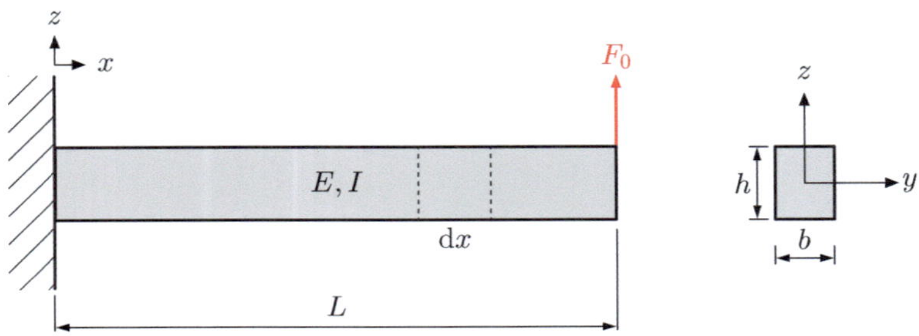

Fig. A.1 General configuration of a cantilever beam with shear force

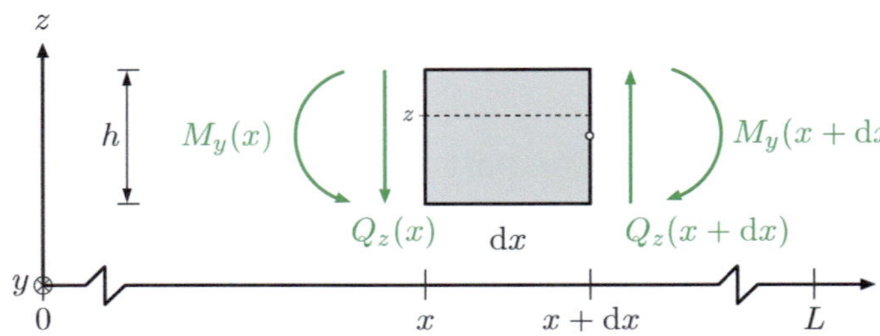

Fig. A.2 Infinitesimal beam element dx in the x-z plane with internal reactions

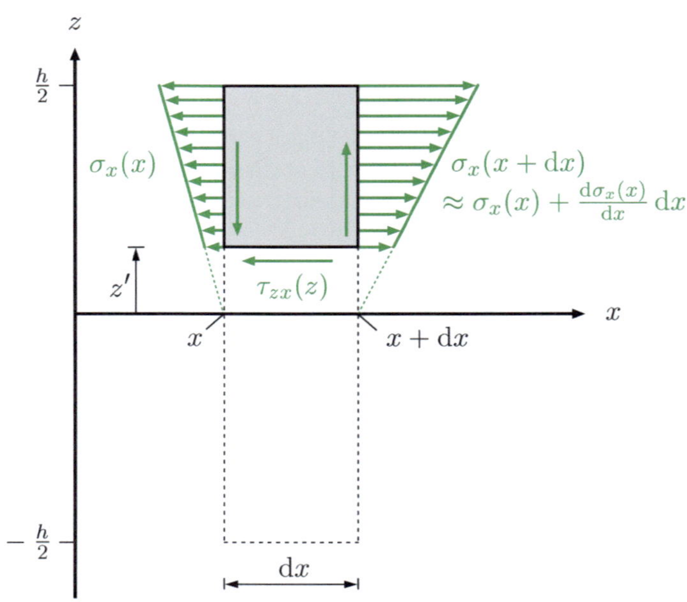

Fig. A.3 Infinitesimal beam element of dimensions d$x \times (h/2 - z')$. The overall configuration is shown in Fig. A.1

or

$$\int \frac{\mathrm{d}\sigma_x(x)}{\mathrm{d}x}\mathrm{d}A - \tau_{zx}(z)b(z) = 0\,. \tag{A.3}$$

Rearranged for

$$\tau_{zx}(z) = \frac{1}{b(z)}\int \frac{\mathrm{d}\sigma_x(x)}{\mathrm{d}x}\mathrm{d}A\,, \tag{A.4}$$

gives with

$$\frac{\mathrm{d}\sigma_x(x)}{\mathrm{d}x} = \frac{\mathrm{d}}{\mathrm{d}x}\left(\frac{M_y(x)}{I_y}\times z\right) = \frac{z}{I_y}\times\frac{\mathrm{d}M_y(x)}{\mathrm{d}x} = \frac{Q_z(x)}{I_y}\times z \tag{A.5}$$

the shear stress distribution

$$\tau_{zx}(z) = \frac{1}{b(z)}\int \frac{Q_z(x)}{I_y}\times z'\mathrm{d}A = \frac{Q_z(x)}{I_y b(z)}\int z'\mathrm{d}A = \frac{Q_z(x)\mathcal{H}_y(z)}{I_y b(z)}, \tag{A.6}$$

where $\mathcal{H}_y(z)$ is the first-order axial moment of area for the part of the cross section according to Fig. A.3. Assuming a rectangular cross section with dimensions $b \times h$, this moment results in:

$$\mathcal{H}_y(z) = \int z'\mathrm{d}A = b\int_z^{h/2} z'\mathrm{d}z' = \frac{b}{2}\left(\frac{h^2}{4} - z^2\right) \tag{A.7}$$

$$= \frac{bh^2}{8}\left[1 - \left(\frac{z}{h/2}\right)^2\right]. \tag{A.8}$$

This gives finally the shear stress distribution for a beam with a rectangular cross section ($b \times h$) and the condition $\tau_{zx} = \tau_{xz}$ to:

$$\tau_{xz}(z) = \frac{Q_z(x)h^2}{8I_y}\left[1 - \left(\frac{z}{h/2}\right)^2\right] \tag{A.9}$$

$$= \frac{3Q_z(x)}{2A}\left[1 - \left(\frac{z}{h/2}\right)^2\right]. \tag{A.10}$$

The maximum of the shear stress distribution results for $z = 0$:

$$\tau_{xz,\max} = \frac{Q_z(x)h^2}{8I_y} = \frac{3Q_z(x)}{2bh} = \frac{3Q_z(x)}{2A}. \tag{A.11}$$

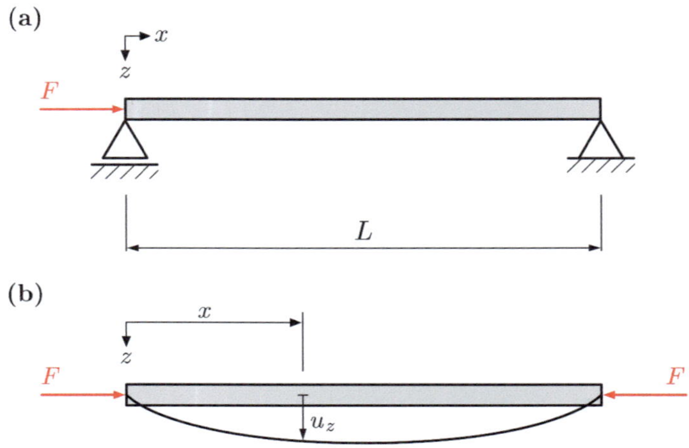

Fig. A.4 Beam hinged on both sides under a compressive load: **a** initial configuration and **b** deformation

Fig. A.5 Cutting free at location x

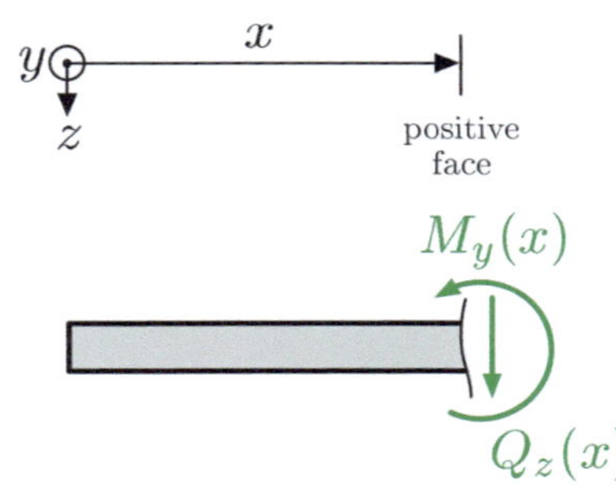

A.1.2 Derivation of the Euler Buckling Force for Homogeneous and Isotropic Euler-Bernoulli Beams

To derive the buckling formula according to Euler, consider a beam hinged on both sides under a compressive force F, see Fig. A.4.

The equilibrium is now established for the first time on the deformed component,[1] see Fig. A.4b.

The equilibrium of moments at the location of the positive face (general position x) yields (see Fig. A.5):

[1] For the derivation of the classical differential equations of Euler-Bernoulli beams under small deformations, the equilibrium is commonly established on the non-deformed component, see Sect. $2.2.1_{\text{TB}}$.

$$\overset{\curvearrowright}{\sum} M_y = 0 \quad \Leftrightarrow \quad +F \times u_z - M_y(x) = 0. \tag{A.12}$$

$$\Rightarrow \quad M_y(x) = +F \times u_z. \tag{A.13}$$

The deformation of an Euler-Bernoulli beam is generally described by a differential equation. For beams with constant bending stiffness EI_y, three classical formulations, based on the distributed load, based on the shear force distribution, or based on the bending moment distribution, can be given, see Eqs. $(2.9)_{\text{TB}}$–$(2.11)_{\text{TB}}$. Using the formulation with the bending moment, we get in our case:

$$EI_y \frac{\mathrm{d}^2 u_z(x)}{\mathrm{d}x^2} = -F \times u_z(x)\,, \tag{A.14}$$

or transformed:

$$EI_y \frac{\mathrm{d}^2 u_z(x)}{\mathrm{d}x^2} + F \times u_z(x) = 0\,. \tag{A.15}$$

Using the abbreviation $\lambda^2 = \frac{F}{EI_y}$ the following representation results:

$$\frac{\mathrm{d}^2 u_z(x)}{\mathrm{d}x^2} + \lambda^2 u_z(x) = 0\,. \tag{A.16}$$

The general solution for such a differential equation is:

$$u_z(x) = c_1 \times \cos(\lambda x) + c_2 \times \sin(\lambda x)\,. \tag{A.17}$$

Using the boundary conditions $u_z(0) = 0$ and $u_z(L) = 0$, the unknown constants c_1 and c_2 can be approached.

From the first boundary condition, i.e. $u_z(0) = 0$, one gets:

$$0 = c_1 \times \cos(0) + c_2 \times \sin(0) \quad \Leftrightarrow \quad c_1 = 0. \tag{A.18}$$

From the second boundary condition, i.e. $u_z(L) = 0$, one gets:

$$0 = c_2 \times \sin(\lambda \cdot L). \tag{A.19}$$

If the product is to be zero, one of the two factors, i.e. c_2 or $\sin(\lambda \times L)$, must be zero. $c_2 = 0$ is a trivial solution (with $c_1 = c_2 = 0$ according to Eq. (A.17): $u_z = 0$, thus no deformation). Therefore, $\sin(\lambda \times L) = 0$ must be looked at more closely. The condition $\sin(\lambda \times L) = 0$ means that $\lambda \times L = k \times \pi$ with $k = 0, 1, 2, \ldots$ (see Fig. A.6).

The condition $\lambda \times L = 0$ would mean $F = 0$ (see definition of λ). This results in the reasonable condition:

$$\lambda \times L = \pi \quad \Leftrightarrow \quad \lambda^2 = \frac{\pi^2}{L^2}. \tag{A.20}$$

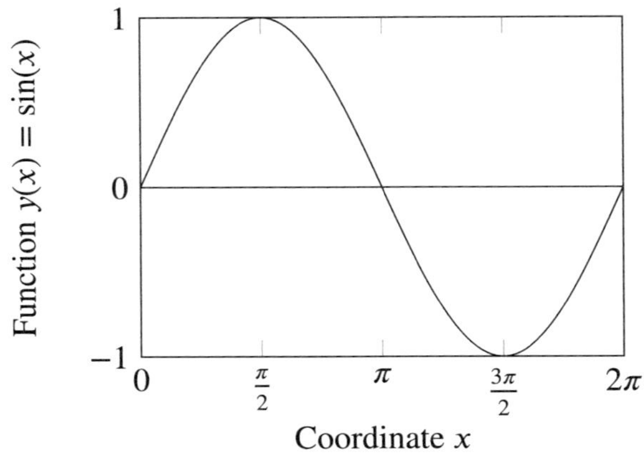

Fig. A.6 Representation of the trigonometric function $\sin(x)$

And finally:

$$F_{\mathrm{cr}} = \frac{\pi^2 E I_{\mathrm{min}}}{L^2}. \tag{A.21}$$

This results in the buckling shape:

$$u_z(x) = c_2 \times \sin\left(\frac{\pi \cdot x}{L}\right), \tag{A.22}$$

where the constant c_2 remains undetermined. It should also be noted that the smaller value of I_y and I_z is to be taken for I_{min}.

Equation (A.21) can also be generalized according to Euler for other support cases by introducing the so-called buckling length L_{cr}. This results in the generalized buckling force according to Euler:

$$F_{\mathrm{cr}} = \frac{\pi^2 E I_{\mathrm{min}}}{L_{\mathrm{cr}}^2}. \tag{A.23}$$

The different formulations of the Euler buckling force are summarized in Table A.1.

Table A.1 Characterization of the classical cases according to Euler

Case	Boundary condition	Buckling length
1	Free—fixed	$L_{\mathrm{cr}} = 2L$
2	Hinged—hinged	$L_{\mathrm{cr}} = L$
3	Hinged—fixed	$L_{\mathrm{cr}} \approx 0.7L$
4	Fixed—fixed	$L_{\mathrm{cr}} = \frac{1}{2}L$

The buckling stress σ_{cr} results from the buckling force using:

$$\sigma_{cr} = \frac{F_{cr}}{A} = \frac{\pi^2 E I_{min}}{A L_{cr}^2}. \tag{A.24}$$

If one defines the so-called slenderness ratio λ by means of the geometric quantities, i.e.

$$\lambda = \frac{L_{cr}}{\sqrt{\frac{I_{min}}{A}}} = L_{cr}\sqrt{\frac{A}{I_{min}}} \quad \Leftrightarrow \quad \lambda^2 = \frac{L_{cr}^2 A}{I_{min}}, \tag{A.25}$$

the buckling stress results in:

$$\sigma_{cr} = \frac{\pi^2 E}{\lambda^2}. \tag{A.26}$$

A.1.3 Newton's Iteration

Newton's method is used for the numerical determination of the roots of functions, see Mitchell (1980). To derive the iteration scheme, a function f at a point x_0 is expanded into a first-order Taylor series:

$$f(x) \approx f(x_0) + f'(x_0) \times (x - x_0) + \cdots. \tag{A.27}$$

Assuming the root at the location x, i.e. $f(x) = 0$, the following iteration rule results (see also Fig. A.7):

$$x_{n+1} = x_n - \frac{f(x_n)}{f'(x_n)}. \tag{A.28}$$

The iteration is performed until the difference between two consecutive abscissa values is less than a given tolerance: $tol = x_{n+1} - x_n$. It should also be noted that the

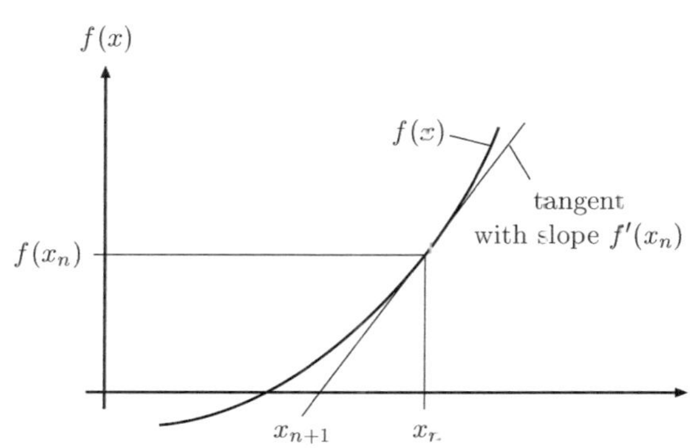

Fig. A.7 Schematic representation of Newton's iteration for the numerical determination of the root of a function

convergence depends on the start value of the iteration. In order to select a meaningful start value, it may be necessary for the function to be represented graphically first.

A.1.4 Numerical Integration of Functions with Variables

When calculating the specific energy absorption SEA in Chap. 4_{TB}, it is possible that difficult integrations have to be carried out. These integrals may contain general variables—such as in the integration in the limits from $x = 0$ to $x = L$—and these can complicate the calculation of the integral. By using computer algebra systems (e.g. Maxima), some of these integrals can be analytically evaluated. However, no solution is found for other integrals with variables and one has to assign concrete numerical values to the variables in order to be able to carry out a *numerical* integration.

Alternatively, a simple numerical scheme should be presented here in order to be able to approximate integrals with variables numerically. For example, consider the following integral:

$$\int_0^L \underbrace{\left(-\left[\frac{x}{L}\right]^2 + \left[\frac{x}{L}\right]^1\right)^{\frac{3}{2}}}_{y(x)} \mathrm{d}x \,. \tag{A.29}$$

This integral can no longer be calculated *analytically* with the computer algebra system Maxima. Let us first consider the integrand, i.e. the function $y(x)$, which is to be integrated, see Fig. A.8a. It is a continuous function and all function values are greater than zero. The integral is represented by the area below the function graph (see the gray area in Fig. A.8a).

For the approximate calculation of the integral, the area under the function graph $y(x)$ is approximated by a series of n rectangles of constant width Δx (see Fig. A.8b), i.e.

$$\Delta x = \frac{L}{n}. \tag{A.30}$$

The abscissa values of the center points of each rectangle result for $1 \leq i \leq n$ in:

$$x_i = (i - 1) \times \Delta x + \frac{\Delta x}{2}. \tag{A.31}$$

Thus, the area A_i of a single rectangle based on the ordinate value $y(x_i) = y_i$ and the interval width is:

$$A_i = \Delta x \times y(x_i) = \Delta x \times y_i \,. \tag{A.32}$$

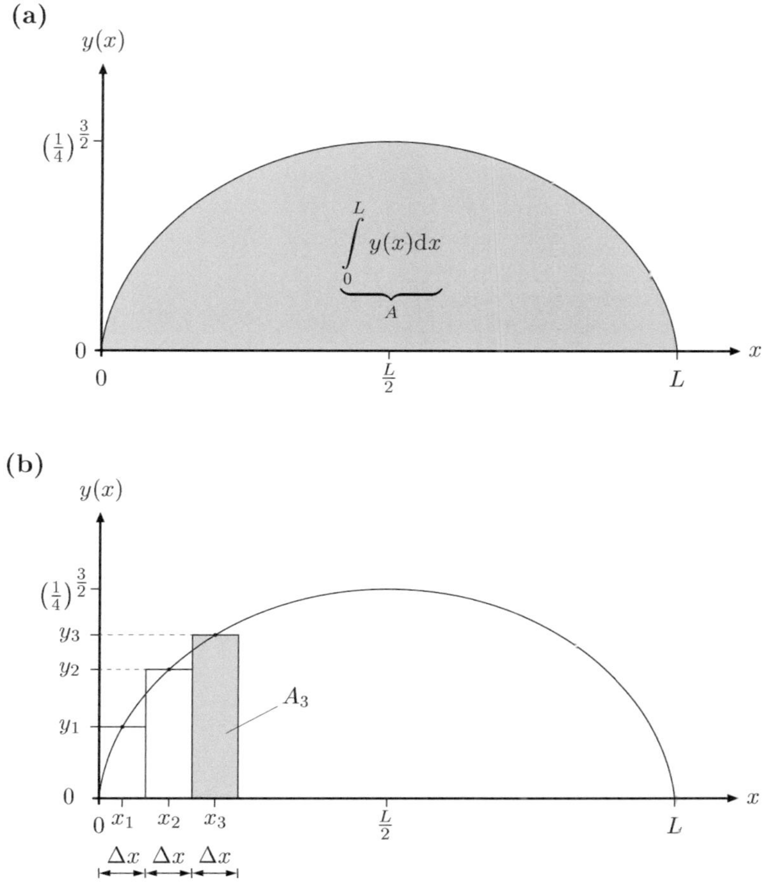

Fig. A.8 Approximation of the integration of a function (with variable), see Eq. (A.29)

Finally, the entire approximation of the area, and thus the integral, results from the summation of the partial areas:

$$\underbrace{\int_0^L y(x)\mathrm{d}x}_{A} \approx \sum_{i=1}^{n} A_i = \sum_{i=1}^{n} \Delta x \times y(x_i)\,. \tag{A.33}$$

The integration scheme according to Eqs. (A.30)–(A.33) can easily be implemented in a computer algebra system (see Fig. A.9) or realized using a classical programming language (see Sect. A.2).

The results of the numerical approximation of the integral according to Eq. (A.29) are summarized in Table A.2. Since no analytical solution to this problem is available,

Numerical Integration of a Function with Variables

```
(%i3)   y : (-(x/L)^2 + (x/L))^(3/2)$

        n_list : [1,2,4,8,16,32,64,128,256,512,1024,2056]$

        for n in n_list do(
            dx : L/n,
            Area : 0,
            for i:1 thru n do(
              current_x : (i-1)*dx + (dx/2),
              current_y : at(y, x=current_x),
              current_A : dx*current_y,
              Area : Area + current_A
            ),
            print("n =",n, "Area =", float(Area)),
            print(" ")
        )$

n = 1 Area = 0.125 L
n = 2 Area = 0.08118988160479111 L
n = 4 Area = 0.07481934508844082 L
n = 8 Area = 0.07382736735303455 L
n = 16 Area = 0.0736645275266041 L
n = 32 Area = 0.07363687899496056 L
n = 64 Area = 0.07363209336908189 L
n = 128 Area = 0.07363125646066039 L
n = 256 Area = 0.07363110932005427 L
n = 512 Area = 0.07363108338032945 L
n = 1024 Area = 0.07363107880109886 L
n = 2056 Area = 0.07363107799047096 L
```

Fig. A.9 Numerical approximation of the integration of a function (with variable) using the computer algebra system Maxima, Makvandi (2020)

Table A.2 Approximation of the integration of a function (with variable), see Eq. (A.29)

n	A	$\|\Delta A\| = \|A_{i+1} - A_i\|$
1	$0.125L$	–
2	$0.08118988160479111L$	$0.04381011839520889L$
4	$0.07481934508844082L$	$0.006370536516350292L$
8	$0.07382736735303455L$	$9.919777354062687 \times 10^{-4}L$
16	$0.0736645275266041L$	$1.628398264304498 \times 10^{-4}L$
32	$0.07363687899496056L$	$2.764853164353986 \times 10^{-5}L$
64	$0.07363209336908189L$	$4.785625878675481 \times 10^{-6}L$
128	$0.07363125646066039L$	$8.369084214948641 \times 10^{-7}L$
256	$0.07363110932005427L$	$1.471406061159808 \times 10^{-7}L$
512	$0.07363108338032945L$	$2.593972482645146 \times 10^{-8}L$
1024	$0.07363107880109886L$	$4.579230591938988 \times 10^{-9}L$
2056	$0.07363107799047096L$	$8.10627898140126 \times 10^{-10}L$

Table A.3 Investigation of the convergence of the integration scheme according to Fig. A.8 based on the function $y(x) = 0$

n	A
1	$5.0L$
2056	$5.0L$
Exact	$5L$

Table A.4 Analysis of the convergence of the integration scheme according to Fig. A.8 based on the function $y(x) = \left(-\left[\frac{x}{L}\right]^2 + \left[\frac{x}{L}\right]^1\right)^{1/2}$

n	A
1	$0.5L$
2	$0.4330127018922193L$
4	$0.4074209160795005L$
8	$0.3979911525764883L$
16	$0.3945858664122887L$
32	$0.3933689759903151L$
64	$0.3929364253352797L$
128	$0.3927330840016931L$
256	$0.3927287966897307L$
512	$0.3927095903102595L$
1024	$0.392702797544798\ L$
2056	$0.3927003878749765L$
4112	$0.3926995435169603L$
Exact	$5\frac{\pi L}{8} = 0.3926990816987241L$

the difference between two approximations was evaluated to check the convergence of the calculation. For the given parameters one recognizes a quite good convergence. Finally, the integration scheme can also be tested on integrals for which there is an analytical solution, see Tables A.3 and A.4. Here, too, one can see that the integrals are approximated with sufficient accuracy from a certain interval width (the exact solution results for the constant function).

A.2 Computer Programs

Factors for Local Instability Calculation

The following Python program B1.py uses Newton's method to calculate the factor B_1 for Eqs. $(6.28)_{\mathrm{TB}}$, $(6.39)_{\mathrm{TB}}$ und $(6.43)_{\mathrm{TB}}$ to be able to evaluate the critical crease stress σ_{cr}. Furthermore, the normalized wavelength $\frac{\lambda}{\Delta h^{\mathrm{C}}}$ is calculated. The program requires a Python 3 installation,[2] which provides the additional libraries sympy and numpy. Using klist, the start and end value or the associated subdivision into steps for the material and geometry parameter k can be specified, see line 8. A list of the Poisson's ratios of the core ν^{C} is specified in vlist, see line 9. By commenting out two of the three equations in the lines 56–58, one can distinguish between the cases in the Sects. $6.1.3_{\mathrm{TB}}$, $6.1.4_{\mathrm{TB}}$ und $6.1.5_{\mathrm{TB}}$. The basics of the Python programming language can be found in Zhang (2015); Padmanabhan (2016); Linge und Langtangen (2016).

B1.py

```
1  import sympy as sp
2  import numpy as np
3  sp.init_printing()
4
5  ###PARAMETERS###
6
7  precision=0.000001
8  klist=np.linspace(0.05,4.00,100)
9  vlist=[0.50]
10 x0=100
11
12 ###EQUATIONS###
13 v,k,x=sp.symbols('v k x')
14
15 expr=(k**2*x**2)/12 + (1/k)*((2/x) * (((3-v)*sp.sinh(x) *
        sp.cosh(x)+(1+v)*x)/((1+v)*(3-v)**2 *(sp.sinh(x))**2
        -((1+v)**3)*x**2)))
16
17 expr2=(k**2*x**2)/12 + (1/k)*((2/x) *
        ((sp.cosh(x)-1)/((1+v)*(3-v)*sp.sinh(x) + ((1+v)**2)*x)))
18
19 expr3=(k**2*x**2)/12 + (1/k)*((2/x) * ((sp.cosh(x)+1)/(3*sp.sinh(x)-x)))
20
21 ###FUNCTION###
22
23
24 def sympy_newton(expr,klist,vlist,precision,x0,label):
25     """Takes a sympy expression, here with parameters v, k, and variable
        x, and lists of parameters k and v, and for a given x0 start value
        and precision calculates the minimum or maximum of function closest
        to x0, for each k and v. The function returns the k, v, wavelength,
        and the y-value of the function at the minimum output is saved in a
        textfile, the name of which is specified with the variable label."""
26     v,k,x=sp.symbols('v k x')
27     first_der=sp.diff(expr, x)
28     second_der=sp.diff(first_der, x)
29     function=sp.lambdify((x,v,k),expr,("numpy", "math", "mpmath",
        "sympy"))
30     function_first_der=sp.lambdify((x,v,k),first_der,("numpy", "math",
        "mpmath", "sympy"))
31     function_second_der=sp.lambdify((x,v,k),second_der,("numpy", "math",
        "mpmath", "sympy"))
```

[2] Thus, the program is called using python3 B1.py.

```
    wavelength=float((2*sp.pi)/x0)
    with open(label+".txt", 'w') as outfile:

     print("{:<16}{:<16}{:<16}{:<20}".format("K","31","2pi/theta","v"),
     file=outfile)
        for k in klist:
            for v in vlist:
                if abs(function_first_der(x0,v,k)) < precision:

     print('{:.6f}'.format(k),'{:.6f}'.format(function(x0,v,k)),'{:.6f}'.
     format(wavelength),'{:.2f}'.format(v), sep='\t', file=outfile)
                else:
                    while abs(function_first_der(x0,v,k)) > precision:
                        if function_second_der(x0,v,k) == 0.0:
                            print("DivByZero error for: ",x0,v,k,
     file=outfile)
                            break
                        x = x0 -
     function_first_der(x0,v,k)/function_second_der(x0,v,k)
                        x0 = x
                        wavelength=float((2*sp.pi)/x0)
                        if abs(function_first_der(x0,v,k)) < precision:

     print('{:.6f}'.format(k),'{:.6f}'.format(function(x0,v,k)),'{:.6f}'.
     format(wavelength),'{:.2f}'.format(v), sep='\t', file=outfile)





###RUN###

sympy_newton(expr,klist,vlist,precision,x0,"case1")
#sympy_newton(expr2,klist,vlist,precision,x0,"case2")
#sympy_newton(expr3,klist,vlist,precision,x0,"case3")
```

Optimal Dimensioning of Sandwich Beams Under Compressive Loads

The following Python program Optm_Comp.py calculates the optimal geometry of a sandwich beam under axial compression using the numerical Newton's method. The output file new.txt contains the coordinates of the point $E\left(\Delta h_E^{\mathrm{C,n}}, \Delta h_E^{\mathrm{F,n}}\right)$ in the Variables h_n_E, t_n_E and $G\left(\Delta h_G^{\mathrm{C,n}}, \Delta h_G^{\mathrm{F,n}}\right)$ in the variables h_n_G, t_n_G. Furthermore, the optimal point is given in the normalized variables h_n, t_n or in absolute values as h, t. At the end, a check is made as to whether the assumption of thin face sheets and a soft core is fulfilled. This program also requires a Python 3 installation, which provides the additional libraries sympy and numpy.

Optm_Comp.py

```
import sympy as sp
import numpy as np
sp.init_printing()

###PARAMETERS###
filename="new"
start=0.03
precision=0.000000000001

b=305.0
L=2540.0
rho_s=2691*9.81*10**(-9)
rho_c=240*9.81*10**(-9)

sig_Y=247.0
E_s=68948.0
E_c=6.8948
G_c=3.4474
P=10*2670
sig_wr=0.5*(E_s*E_c*G_c)**(1/3)

if sig_wr < sig_Y:
    sig_limit=sig_wr
else:
    sig_limit=sig_Y

###CALCULATION_COORDINATES_POINT_E###
h_n_E=np.sqrt((P/(2.0*b*L*G_c))**2.0 + 4.0*sig_limit/(np.pi**2.0*E_s)) + P/(2.0*b*L*G_c)

t_n_E=P/(2.0*b*L*sig_limit)


###EQUATION_FOR_NEWTON_ITER###
x = sp.symbols('x')
f=((4*rho_s/(np.pi**2*E_s))/(x*((x*b*L/P)-1/G_c)))+rho_c*x


###CALCULATION_COORDINATES_POINT_G###

def sympy_newton(expr,precision,x0,label):
    """Takes a sympy expression with variable x, and for a given x0 start value and precision calculates the minimum or maximum of function closest to x0."""
    x=sp.symbols('x')
    first_der=sp.diff(expr, x)
    second_der=sp.diff(first_der, x)
    function=sp.lambdify((x),expr,("numpy", "math", "mpmath", "sympy"))
    function_first_der=sp.lambdify((x),first_der,("numpy", "math", "mpmath", "sympy"))
    function_second_der=sp.lambdify((x),second_der,("numpy", "math", "mpmath", "sympy"))
    iter=0
    with open(label+".txt", 'w') as outfile:
        if abs(function_first_der(x0)) < precision:
```

```
            return x0
        else:
            while abs(function_first_der(x0)) > precision:
                if function_second_der(x0) == 0:
                    print("DivByZero error for: ",x0, file=outfile)
                    break
                x = x0 - function_first_der(x0)/function_second_der(x0)
                x0 = x
                iter+=1
                if abs(function_first_der(x0)) < precision:
                    return x0
                if iter > 50:
                    print("Bad starting value or precision, over 50 iterations", file=outfile)
                    break



###RUN###
h_n_G=sympy_newton(f,precision,start,filename)


if h_n_G is not None: #(check is Newton iteration was successfully completed)
    f_t_n_G=(((2*P)/(np.pi**2*b*L*E_s))/(x*(x-P/(G_c*b*L))))
    l_t_n_G=sp.lambdify((x), f_t_n_G, ("numpy", "math", "mpmath", "sympy"))

    t_n_G=l_t_n_G(h_n_G)


    ###OUTPUT###

    with open(filename+".txt", 'a') as outfile:
        print("h_n_G = ",h_n_G, file=outfile)
        print("t_n_G = ",t_n_G, file=outfile)
        print("h_n_E = ",h_n_E, file=outfile)
        print("t_n_E = ",t_n_E, file=outfile)

        if h_n_G <= h_n_E:
            h_n=h_n_G
            t_n=t_n_G
        else:
            h_n=h_n_E
            t_n=t_n_E
        print("h_n = ",h_n, file=outfile)
        print("t_n = ",t_n, file=outfile)


        h=h_n*L
        t=t_n*L
        print("h = ",h, file=outfile)
        print("t = ",t, file=outfile)

    ###CHECK_SANDWICH_THIN_WEAK_CORE###
```

```
    ch_wt1=(6*E_s*t*h**2)/(E_c*h**3)
    #print(ch_wt1)
    ch_wt2=h/t
    #print(ch_wt2)

    with open(filename+".txt", 'a') as outfile:
        if (ch_wt1 >= 100) and (ch_wt2 >=4.77):
            print("Sandwich with WEAK core and THIN faces", file=outfile)
        else:
            print("NO weak core and thin faces", file=outfile)
```

Optimal Dimensioning of Sandwich Beams Under Bending Loads

The following Python program Optm_Comp_Bend_1.py calculates the optimal geometry of a sandwich beam under bending loads using the numerical Newton's method. Furthermore, the case of a simply supported beam with a central point force (see Fig. 5.19$_{\text{TB}}$) is considered. The output file test3.txt contains the coordinates of the point $E\left(\Delta h_E^{\text{C,n}}, \Delta h_E^{\text{F,n}}\right)$ in the Variables h_n_E, t_n_E, $A\left(\Delta h_A^{\text{C,n}}, \Delta h_A^{\text{F,n}}\right)$ in the variables h_n_A, t_n_A, $D\left(\Delta h_D^{\text{C,n}}, \Delta h_D^{\text{F,n}}\right)$ in the variables h_n_D, t_n_D, $B\left(\Delta h_B^{\text{C,n}}, \Delta h_B^{\text{F, n}}\right)$ in the variables h_n_B, t_n_B and $C\left(\Delta h_C^{\text{C,n}}, \Delta h_C^{\text{F,n}}\right)$ in the variables h_n_C, t_n_C. Furthermore, the second pole of the function g_3 is provided as variable g3 pole. This program also requires a Python 3 installation, which provides the additional sympy and numpy libraries.

Optm_Comp_Bend_1.py

```
import sympy as sp
import numpy as np
sp.init_printing()

###PARAMETERS###
filename="test3"
start=0.085
precision=0.000000000001

b=305.0
L=2540.0
r_1=0.003

rho_s=2691*9.81*10**(-9)
rho_c=240*9.81*10**(-9)

sig_Y=247.0
E_s=68948.0
E_c=6.8948
G_c=3.4474
tau=E_c/50.0
```

```

F=2667.0
Q_max=F/2.0
M_max=(F*L)/4.0

P=10*2670
sig_wr=0.5*(E_s*E_c*G_c)**(1/3)

if sig_wr < sig_Y:
    sig_limit=sig_wr
else:
    sig_limit=sig_Y

###CALCULATION_COORDINATES_POINT_E###
h_n_E=Q_max/(b*L*tau)
t_n_E=(M_max*tau)/(Q_max*L*sig_limit)

with open(filename+".txt", 'a') as outfile:
    print("Point E: h_n_E , t_n_E= ",h_n_E,", ",t_n_E,
    file=outfile)


###CALCULATION_COORDINATES_POINT_A###
h_n_A=(F/(M_max*r_1)*(2*L*sig_limit/(48*E_s)+M_max/(4*b*L*G_c)))
t_n_A=M_max/(b*L**2*sig_limit*h_n_A)
with open(filename+".txt", 'a') as outfile:
    print("Point A: h_n_A , t_n_A= ",h_n_A,", ",t_n_A,
    file=outfile)


###CALCULATION_COORDINATES_POINT_D###
h_n_D=Q_max/(b*L*tau)
t_n_D=2*b*L*F*tau**2/(48*E_s*Q_max**2*(r_1-F*tau/(4*G_c*Q_max)))

with open(filename+".txt", 'a') as outfile:
    print("Point D: h_n_D , t_n_D= ",h_n_D,", ",t_n_D,
    file=outfile)


###CALCULATION_COORDINATES_POINT_B###
h_n_B=np.sqrt(2*rho_s*M_max/(b*L**2*sig_limit*rho_c))
t_n_B=M_max/(b*L**2*sig_limit*h_n_B)

with open(filename+".txt", 'a') as outfile:
    print("Point B: h_n_B , t_n_B= ",h_n_B,", ",t_n_B,
    file=outfile)


###CALCULATION_COORDINATES_POINT_C###
###EQUATION_FOR_NEWTON_ITER###
x = sp.symbols('x')
f=(2*rho_s*2*F/(48*E_s*b*L*x**2))/(r_1-F/(4*G_c*b*L*x))+rho_c*x


```

```
def sympy_newton(expr,precision,x0,label):
    """Takes a sympy expression with variable x, and for a given x0 start value and precision calculates the minimum or maximum of function closest to x0. The output is saved in a textfile, the name of which is specified with the variable label."""
    x=sp.symbols('x')
    first_der=sp.diff(expr, x)
    second_der=sp.diff(first_der, x)
    function=sp.lambdify((x),expr,("numpy", "math", "mpmath", "sympy"))
    function_first_der=sp.lambdify((x),first_der,("numpy", "math", "mpmath", "sympy"))
    function_second_der=sp.lambdify((x),second_der,("numpy", "math", "mpmath", "sympy"))
    iteration=0
    with open(label+".txt", 'a') as outfile:
        if abs(function_first_der(x0)) < precision:
            return x0
        else:
            while abs(function_first_der(x0)) > precision:
                if function_second_der(x0) == 0:
                    print("ZeroDivisionError for: ",x0," at iteration ",iteration, file=outfile)
                    break
                x = x0 - function_first_der(x0)/function_second_der(x0)
                x0 = x
                iteration+=1
                if abs(function_first_der(x0)) < precision:
                    return x0
                if iteration > 50:
                    print("Bad starting value or precision, over 50 iterations", file=outfile)
                    break


###RUN###
h_n_C=sympy_newton(f,precision,start,filename)
t_n_C=(2*F/(48*E_s*b*L*h_n_C**2))/(r_1-F/(4*G_c*b*L*h_n_C))

with open(filename+".txt", 'a') as outfile:
    print("Point C: h_n_C , t_n_C= ",h_n_C,", ",t_n_C, file=outfile)


###CALCULATION_POLE_G3###

g3_pole=F/(4*G_c*b*L*r_1)
print(g3_pole)

with open(filename+".txt", 'a') as outfile:
    print("g3 pole= ",g3_pole, file=outfile)
```

Numerical Integration of Functions with Variables

The following Python program numerical.py approximately calculates an integral, whereby the area under the function graph $y(x)$ is approximated by a series of n rectangles of constant width Δx (see Sect. A.1.4 for details).

numerical.py

```
import sympy as sp
import numpy as np
sp.init_printing()

x=sp.symbols('x')
L=sp.symbols('L')
y = (-(x/L)**(2) + (x/L))**(3/2)
n_list = [1,2,4,8,16,32,64,128,256,512,1024,2056]
for n in n_list:
    dx = np.divide(L,n)
    area = 0
    for i in range(1, n+1):
        current_x = (i-1)*dx + (dx/2)
        current_y = sp.lambdify((x),y)
        current_A = dx*current_y(current_x)
        area = area + current_A
    print('n =',n, 'Area =', area)
```

The result of executing the program numerical.py is shown in Fig. A.10.

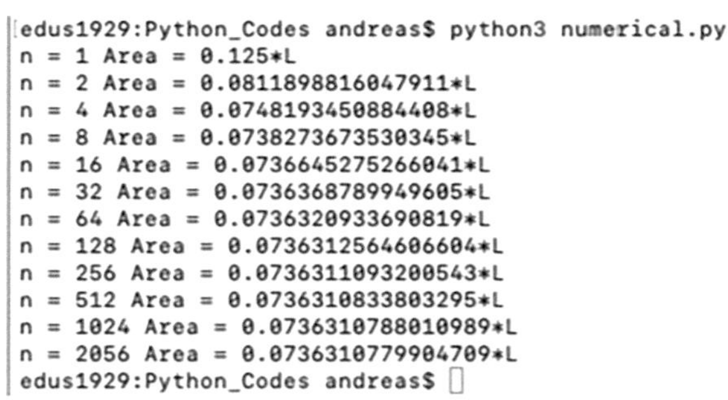

```
edus1929:Python_Codes andreas$ python3 numerical.py
n = 1 Area = 0.125*L
n = 2 Area = 0.0811898816047911*L
n = 4 Area = 0.0748193450884408*L
n = 8 Area = 0.0738273673530345*L
n = 16 Area = 0.0736645275266041*L
n = 32 Area = 0.0736368789949605*L
n = 64 Area = 0.0736320933690819*L
n = 128 Area = 0.0736312564606604*L
n = 256 Area = 0.0736311093200543*L
n = 512 Area = 0.0736310833803295*L
n = 1024 Area = 0.0736310788010989*L
n = 2056 Area = 0.0736310779904709*L
edus1929:Python_Codes andreas$
```

Fig. A.10 Numerical calculation of an integral, see program numerical.py

References

Linge, S., Langtangen, H.P.: Programming for Computations - Python: A Gentle Introduction to Numerical Simulations with Python. Springer, Cham (2016)
Makvandi, R.: Private Kommunikation. Institute of Mechanics, Otto von Guericke University Magdeburg, Germany (2020)
Mitchell, A.R., Griffiths, D.F.: The Finite Difference Method in Partial Differential Equations. Wiley, New York (1980)
Padmanabhan, T.R.: Programming with Python. Springer, Singapore (2016)
Zhang, Y.: An Introduction to Python and Computer Programming. Springer, Singapore (2015)

Index

A. Öchsner, *Exercises and Solutions in Lightweight Design*,
https://doi.org/10.1007/978-3-658-49333-2

MIX
Papier aus verantwortungsvollen Quellen
Paper from responsible sources
FSC® C105338

If you have any concerns about our products,
you can contact us on
ProductSafety@springernature.com

In case Publisher is established outside the EU,
the EU authorized representative is:
Springer Nature Customer Service Center GmbH
Europaplatz 3, 69115 Heidelberg, Germany

Printed by Libri Plureos GmbH
in Hamburg, Germany